HSC Year 12
CHEMISTRY

JAYA CHOWDHURY | SAMANTHA DREON

NESA 2017 SYLLABUS + 2021 EXAM QUESTIONS

A+

+ 13 topic tests
+ two complete practice exams
+ detailed sample answers

PRACTICE
EXAMS

Nelson

A+ HSC Chemistry Practice Exams
1st Edition
Jaya Chowdhury
Samantha Dreon
ISBN 9780170465274

Publisher: Alice Wilson
Series editor: Catherine Greenwood
Series text design: Nikita Bansal
Series cover design: Nikita Bansal
Series designer: Cengage Creative Studio
Reviewer: Col Harrison
Artwork: MPS Limited
Production controller: Karen Young
Typeset by: Nikki M Group Pty Ltd

Any URLs contained in this publication were checked for currency during the
production process. Note, however, that the publisher cannot vouch for the
ongoing currency of URLs.

For product information and technology assistance,
in Australia call **1300 790 853**;
in New Zealand call **0800 449 725**

For permission to use material from this text or product, please email
aust.permissions@cengage.com

ISBN 978 0 17 046527 4

Cengage Learning Australia
Level 7, 80 Dorcas Street
South Melbourne, Victoria Australia 3205

Cengage Learning New Zealand
Unit 4B Rosedale Office Park
331 Rosedale Road, Albany, North Shore 0632, NZ

For learning solutions, visit **cengage.com.au**

Printed in China by 1010 Printing International Limited.
1 2 3 4 5 6 7 26 25 24 23 22

CONTENTS

CHAPTER 1

Module 5: Equilibrium and acid reactions

CHAPTER 2

Module 6: Acid/base reactions

CHAPTER 3

Module 7: Organic chemistry

CHAPTER 4

Module 8: Applying chemical ideas

Practice exams

9780170465274

HOW TO USE THIS BOOK

The *A+ HSC Chemistry* resources are designed to be used year-round to prepare you for your HSC Chemistry exam. *A+ HSC Chemistry Practice Exams* includes 13 topic tests and two practice exams, plus detailed solutions for all questions. This section gives you a brief overview of the features included in this resource.

Topic tests

Each topic test addresses one inquiry question of Modules 5–8 of the syllabus. The tests follow the same sequence as the syllabus, starting with the first inquiry question of Module 5 and ending with the final inquiry question of Module 8. Each topic test includes multiple-choice and short-answer questions.

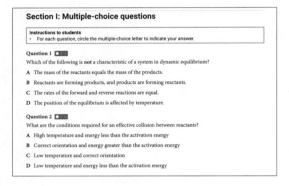
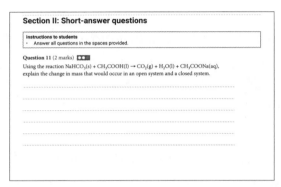

Practice exams

Both practice exams cover all content from Modules 5–8 of the HSC Chemistry syllabus. The practice exams have perforated pages so that you can remove them from the book and practise under exam-style conditions.

Solutions

Solutions to topic tests and practice exams are supplied at the back of the book. They have been written to reflect a high-scoring response and include explanations of what makes an effective answer.

Explanations

The solutions section includes explanations of each multiple-choice option, both correct and incorrect. Explanations of written response items explain what a high-scoring response looks like and signposts potential mistakes.

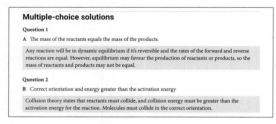
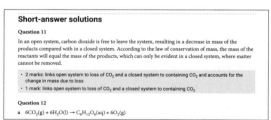

Icons

You will notice the following icons in the topic tests and practice exams.

©NESA | 2020 SI Q17

This icon appears with official past NESA questions.

These icons indicate whether the question is easy, medium or hard.

A+ HSC Chemistry Study Notes

A+ HSC Chemistry Practice Exams can be used independently\, or alongside the accompanying resource *A+ HSC Chemistry Study Notes*. *A+ HSC Chemistry Study Notes* includes topic summaries and exam practice for all key knowledge in the HSC Chemistry syllabus that you will be assessed on during the exam, as well as detailed revision and exam preparation advice to help you get ready for the exam.

A+ DIGITAL

Just scan the QR code or type the URL into your browser to access:

*   A+ Flashcards: revise key terms and concepts online

*   Revision summaries of all concepts from each inquiry question.

Note: You will need to create a free NelsonNet account.

https://get.ga/
aplus-hsc-chemistry-u34

ABOUT THE AUTHORS

Jaya Chowdhury

Jaya Chowdhury is a passionate educator with more than 25 years of teaching experience in Western Australia and New South Wales. Jaya is an academic at Macquarie University in the Department of Educational Studies. She delivers workshops for Masters of Teaching and final year secondary science education students. Jaya also assists the Widening Participation team at Macquarie University. Jaya was the HSC Content Manager at Access Macquarie Pty Ltd and guided the development of online resources for the HSC courses, including the new NSW Chemistry syllabus.

Samantha Dreon

Samantha Dreon began her career as a food and pharmaceutical quality control technician. For the past 10 years, she has been teaching Science and Chemistry in Marist schools across Sydney, refining the pedagogical practice of the Flipped Classroom. Most recently, Sam has been acting as an Expert Teacher for Sydney Catholic Schools, programming Junior and Senior Chemistry curriculums.

CHAPTER 1
MODULE 5: EQUILIBRIUM AND ACID REACTIONS

Test 1: **Static and dynamic equilibrium and factors that affect equilibrium**

Section I: 10 marks. Section II: 25 marks. Total marks: 35.
Suggested time: 63 minutes

Section I: Multiple-choice questions

Instructions to students
- For each question, circle the multiple-choice letter to indicate your answer.

Question 1

Which of the following is **not** a characteristic of a system in dynamic equilibrium?

A The mass of the reactants equals the mass of the products.

B Reactants are forming products, and products are forming reactants.

C The rates of the forward and reverse reactions are equal.

D The position of the equilibrium is affected by temperature.

Question 2

What are the conditions required for an effective collision between reactants?

A High temperature and energy less than the activation energy

B Correct orientation and energy greater than the activation energy

C Low temperature and correct orientation

D Low temperature and energy less than the activation energy

Question 3

Which of the following changes will shift this equilibrium to the right?

$$H_2(g) + I_2(g) \rightleftharpoons 2HI(g) \qquad \Delta H = -9.4\,\text{kJ}\,\text{mol}^{-1}$$

A Increasing the pressure

B Adding a catalyst

C Increasing the temperature

D Increasing the concentration of I_2

Question 4

Energy profile diagrams for two different chemical reactions (A and B) are shown.

Which of these reactions is more likely to be reversible and why?

A Reaction A, because its forward reaction is endothermic

B Reaction B, because its forward reaction is exothermic

C Reaction A, because the activation energy for its reverse reaction is less than that for reaction B

D Neither, because the activation energies for their forward reactions are the same

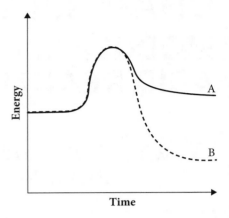

Question 5

The Haber process is used to produce ammonia for the manufacture of plastics, explosives and fertilisers. The equation of the Haber process is:

$$N_2(g) + 3H_2(g) \rightleftharpoons 2NH_3(g) \qquad \Delta H = +92 \, kJ \, L^{-1}$$

What is the effect of increasing the temperatures on the rates of the forward and reverse reactions in the Haber process?

A The rates of the forward and reverse reactions increase equally.

B The rate of the forward reaction remains unchanged, whereas the rate of the reverse reaction increases.

C The rates of both reactions increase, but the rate of the reverse reaction increases more than the rate of the forward reaction.

D The rates of both reactions increase, but the rate of the forward reaction increases more than the rate of the reverse reaction.

Question 6

A method used by Indigenous Australians to detoxify certain seeds is to crush and soak the seeds in water, removing the water-soluble toxins from the seed. An energy profile diagram of this process is shown below.

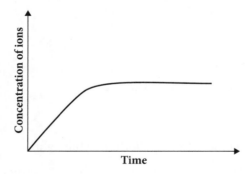

Which statement correctly accounts for the shape of this graph?

A The concentration of ions increases as the seeds are crushed.

B The concentration of ions increases until equilibrium is established with the water.

C The concentration of ions increases until the toxin has been completely removed.

D The concentration of ions increases the longer the seeds soak in the water.

9780170465274

Question 7 ⬤⬤⬤

During your Chemistry course, you studied an equilibrium system involving nitrogen dioxide and dinitrogen tetroxide. A sealed glass vial contains nitrogen dioxide and dinitrogen tetroxide in equilibrium, as represented by the following equation:

$$2NO_2(g) \rightleftharpoons N_2O_4(g) \qquad \Delta H < 0$$

At 25°C, the vial has a pale brown colour. What colour change, if any, would occur if the glass vial was heated to 80°C?

A There would be no noticeable change in colour.

B It would become darker brown.

C It would become colourless.

D It would become pale, and then return to a pale brown colour as equilibrium was restored.

Question 8 ⬤⬤⬤

Which of the following energy profile diagrams best represents a spontaneous endothermic reaction?

A
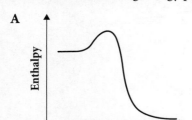

B
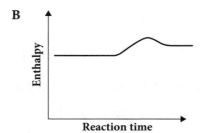

C
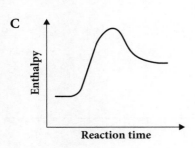

D
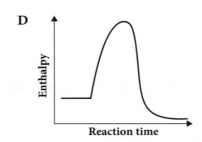

Question 9 ⬤⬤⬤

Cobalt chloride hexahydrate is part of the following equilibrium in solution at 25°C:

$$Co(H_2O)_6{}^{2+}(aq) + 4Cl^-(aq) \rightleftharpoons CoCl_4{}^{2-}(aq) + 6H_2O(aq)$$

What would occur if sodium chloride solution was added?

A Sodium chloride would have no impact on the equilibrium.

B The concentration of $CoCl_4{}^{2-}$ would increase.

C The colour of the solution would change from blue to pink.

D The concentration of Cl^- would decrease.

Question 10 ©NESA 2014 SI Q20 ●●●

This graph represents the yield of an equilibrium reaction at different temperature and pressure conditions inside a reaction vessel.

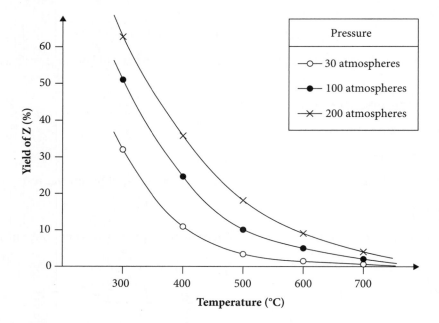

Which of the following reactions could produce the trends shown in the graph?

A $X(g) + Y(g) \rightleftharpoons 3Z(g)$ $\Delta H = +100\,kJ$

B $X(g) + Y(g) \rightleftharpoons 2Z(g)$ $\Delta H = -100\,kJ$

C $2X(g) + 2Y(g) \rightleftharpoons Z(g)$ $\Delta H = +100\,kJ$

D $4X(g) + 2Y(g) \rightleftharpoons 3Z(g)$ $\Delta H = -100\,kJ$

Section II: Short-answer questions

Instructions to students
- Answer all questions in the spaces provided.

Question 11 (2 marks) ●●

Using the reaction $NaHCO_3(s) + CH_3COOH(l) \rightarrow CO_2(g) + H_2O(l) + CH_3COONa(aq)$, explain the change in mass that would occur in an open system and a closed system.

Question 12 (9 marks)

A student recorded the following data in a recent experiment on photosynthesis.

Species	$CH_4(g)$	$O_2(g)$	$CO_2(g)$	$H_2O(l)$	$C_8H_{18}(l)$	$C_6H_{12}O_6(s)$
Enthalpy of formation (kJ mol^{-1})	+75	0	−381	−287	−250	−1289
Entropy (J mol^{-1})	+186	+187	+236	+54	+361	+234

a ◐ Write the balanced chemical equation for photosynthesis.   1 mark

_____

b ◑ Using the data supplied above, calculate the ΔG of photosynthesis, assuming standard conditions.   5 marks

_____

_____

_____

_____

_____

_____

_____

_____

_____

c ● Using the values you calculated in part **b**, analyse the effects of entropy and enthalpy on photosynthesis.   3 marks

_____

_____

_____

_____

_____

_____

_____

Question 13 (6 marks)

The graph shows the changes occurring in an equilibrium system in which N_2O_4 and NO_2 react according to the equation:

$$N_2O_4(g) \rightleftharpoons 2NO_2(g) \qquad \Delta H = +58\,kJ\,mol^{-1}$$

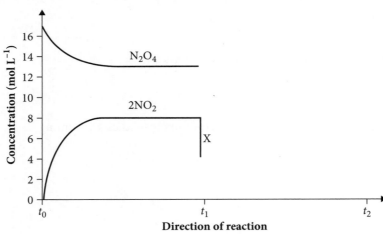

a ⬤◖◗ Identify the change made to the system at time t_1 indicated by the X on the graph.   1 mark

_____

b ◖⬤◗ Complete the graph, showing the system returning to equilibrium by t_2.   2 marks

c ◖⬤◗ After equilibrium was re-established, the volume of the reaction vessel was decreased.

Explain, using collision theory, how this change would have affected the equilibrium
for this system.   3 marks

_____

_____

_____

_____

_____

_____

Question 14 (4 marks) ⬤⬤⬤

The formation of hydrogen chloride gas is represented by the following energy profile.

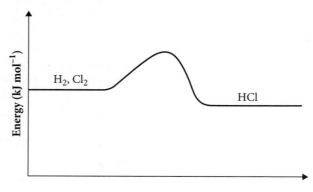

Use the energy profile to explain how an increase in temperature would affect the activation energy and the rate of reaction.

_____

_____

_____

_____

_____

_____

_____

_____

_____

_____

Question 15 (4 marks)

Within the closed system of a soft-drink can, the following chemical reaction occurs:

$$H_2O(l) + CO_2(g) \rightleftharpoons H_2CO_3(aq)$$

When a can of soft drink is opened, carbon dioxide gas leaves the system. As a result, carbonic acid decomposes to re-establish equilibrium. The can of soft drink will go 'flat' if it is left open for an extended period of time.

To speed up this process, a student added a solution of concentrated sodium hydroxide to the can.

a　◖●○○◗　Write a balanced chemical equation for this reaction.　　1 mark

_____

b　◖●●●◗　Explain, using Le Chatelier's principle, how the addition of sodium hydroxide reduces the time it takes for a can of soft drink to go flat.　　3 marks

_____

_____

_____

_____

_____

_____

Test 2:   Calculating the equilibrium constant (K_{eq})

Section I: 10 marks. Section II: 25 marks. Total marks: 35.
Suggested time: 60 minutes

Section I: Multiple-choice questions

Instructions to students
- For each question, circle the multiple-choice letter to indicate your answer.

Question 1

What is the correct definition of a homogeneous equilibrium?

A   All species are in the same state.

B   All species have the same concentration.

C   All species have the same molar ratio.

D   All species have the same rate of reaction.

Question 2

Consider the equilibrium reaction:

$$2HI(g) \rightleftharpoons H_2(g) + I_2(g)$$

Identify the correct equilibrium expression.

A $\dfrac{[2HI]}{[H_2][I_2]}$

B $\dfrac{[HI]^2}{[H_2][I_2]}$

C $\dfrac{[H_2][I_2]}{[HI]^2}$

D $\dfrac{[H_2][I_2]}{[2HI]}$

Question 3

Which row in the table correctly describes what happens when the concentration of reactant is increased within a system at equilibrium?

	K_{eq}	Reaction favours:
A	Decreases	Product
B	Remains unchanged	Product
C	Decreases	Reactant
D	Remains unchanged	Reactant

Question 4 ⬤⬤

The formation of nitrogen trifluoride is described by the following equation:

$$N_2(g) + 3F_2(g) \rightleftharpoons 2NF_3(g) \qquad \Delta H = -264\,\text{kJ mol}^{-1}$$

Under which conditions would the concentration of nitrogen trifluoride increase?

A Decrease in pressure, increase in temperature

B Decrease in pressure, decrease in temperature

C Increase in pressure, increase in temperature

D Increase in pressure, decrease in temperature

Question 5 ⬤

The graph shows the concentrations over time for the system:

$$2NO(g) + O_2(g) \rightleftharpoons 2NO_2(g) \qquad \Delta H = -62\,\text{kJ mol}^{-1}$$

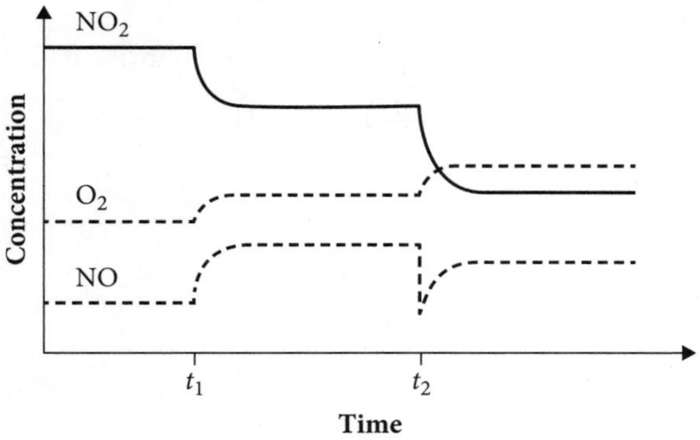

What has happened at times t_1 and t_2?

	t_1	t_2
A	Temperature increased	NO removed
B	Temperature decreased	O_2 added
C	Pressure increased	O_2 added
D	Pressure decreased	NO removed

Question 6 ⬤⬤⬤

Consider the following reaction systems at equilibrium:

$$\text{System 1: } CO(g) + 3H_2(g) \rightleftharpoons CH_4(g) + H_2O(g) \qquad K_{eq} = 9.17 \times 10^{-2}$$

$$\text{System 2: } CH_4(g) + 2H_2S(g) \rightleftharpoons CS_2(g) + 4H_2(g) \qquad K_{eq} = 3.3 \times 10^{4}$$

Which of these statements regarding these systems is/are true?

  i System 2 reaches equilibrium faster than system 1.

 ii The greatest ratio of products to reactants occurs in system 2.

iii Equilibrium for system 1 favours the reactants more than it does for system 2.

A i only

B ii only

C i and iii only

D ii and iii only

Question 7 〔◯◯■〕

Consider the equilibrium for the following equation:

$$P_4O_{10}(s) \rightleftharpoons P_4(s) + 5O_2(g) \qquad \Delta H < 0$$

Which of the following changes would cause the magnitude of the equilibrium constant for this reaction to increase?

A The temperature is decreased.

B The pressure is increased.

C The mass of P_4O_{10} is increased.

D The concentration of O_2 is increased.

Question 8 〔◯◯◯〕

At 400°C, the equilibrium constant is 20 for the following reaction:

$$2N_2O_5(g) \rightleftharpoons 4NO_2(g) + O_2(g)$$

What is the equilibrium constant for the following reaction at equilibrium?

$$2NO_2(g) + \tfrac{1}{2}O_2(g) \rightleftharpoons N_2O_5(g)$$

A 0.05

B 0.22

C 10

D 20

Question 9 〔◯◯◯〕

Determine the K_{eq} for the following reaction if 14 g of iron sulfide was added to a 1 L reaction vessel with $1.00\,\text{mol}\,L^{-1}$ oxygen. At equilibrium, 0.233 mol of sulfur dioxide is produced:

$$4FeS_2(s) + 11O_2(g) \rightleftharpoons 2Fe_2O_3(s) + 8SO_2$$

A 2.90×10^8

B 1.66×10^3

C 6.08×10^{-4}

D 3.42×10^{-1}

Question 10 〔◯■■〕

A base 'B–OH' has a base dissociation constant, K_b, in aqueous solution.

What is the equilibrium expression?

A $[B^+][OH^-]$

B $[BOH][B^+]$

C $\dfrac{K_b}{K_w}$

D $\dfrac{[B^+][OH^-]}{[BOH]}$

Section II: Short-answer questions

> **Instructions to students**
> • Answer all questions in the spaces provided.

Question 11 (9 marks)

Chemical engineers clean semiconductors by adding chlorine trifluoride. When it decomposes, chlorine trifluoride reacts with the semiconductor material, allowing for it to be cleaned without the need to dismantle the equipment. Chlorine trifluoride decomposes according to the following equation:

$$2ClF_3(g) \rightleftharpoons 3F_2(g) + Cl_2(g)$$

a ⬤⬤ During cleaning, the concentration of both chlorine trifluoride and chlorine gas at equilibrium was $2.5 \, mol \, L^{-1}$ at 15°C.

Determine the initial concentration of chlorine trifluoride.   3 marks

b ⬤ Write the equilibrium expression for this reaction.   1 mark

c ⬤⬤ Calculate the K_{eq} for this reaction.   2 marks

d ⬤⬤ This decomposition reaction is endothermic.

Write the equilibrium expression for the exothermic reaction.   1 mark

e 〔OO 〕 Determine the K_{eq} for the exothermic reaction. 2 marks

_____

_____

_____

_____

Question 12 (7 marks)

a 〔OO 〕 Consider the reaction:

$$CH_4(g) + 2H_2S(g) \rightleftharpoons CS_2(g) + 4H_2(g) \qquad K_{eq} = 3.65 \text{ at } 900°C$$

At this temperature, 1.00 mol of CH_4, 2.00 mol of H_2S, 1.00 mol of CS_2 and 2.00 mol of H_2 are mixed in a 200 mL vessel.

Calculate the reaction quotient, Q_{eq}, and predict the direction in which the reaction will proceed to reach equilibrium. 3 marks

_____

_____

_____

_____

_____

_____

_____

b 〔OOO 〕 Show that the system is at equilibrium when the concentration of methane is 6.14 mol L^{-1}. 4 marks

_____

_____

_____

_____

_____

_____

_____

_____

Question 13 (4 marks) ⬤⬤

Hydrogen fluoride decomposes to form hydrogen and fluorine gases:

$$2HF(g) \rightleftharpoons H_2(g) + F_2(g)$$

In an experiment, a chemist recorded the concentration of hydrogen at equilibrium at several temperatures and presented the data in the graph below.

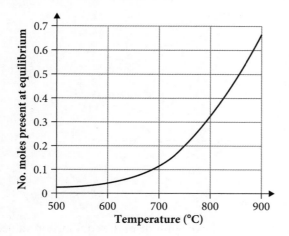

Predict the effect an increase in temperature will have on the K_{eq} value. Justify your prediction.

_____

_____

_____

_____

_____

_____

_____

_____

Question 14 (5 marks)

The graph shows concentration changes that occur in a closed system.

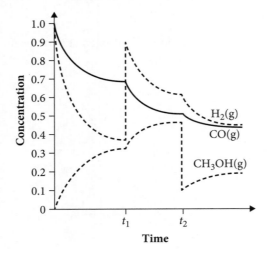

a At t_1, a change was made to the system.

Determine the impact this change had on the magnitude of the equilibrium value.   4 marks

_____

_____

_____

_____

_____

_____

_____

_____

b The new temperature, and therefore the new equilibrium position, was maintained. At t_2, CH_3OH was removed from the system. Predict what effect, if any, this change would have had on the magnitude of the equilibrium value.   1 mark

_____

_____

_____

Test 3:   Solution equilibria

Section I: 10 marks. Section II: 24 marks. Total marks: 34.
Suggested time: 60 minutes

Section I: Multiple-choice questions

Instructions to students
- For each question, circle the multiple-choice letter to indicate your answer.

Question 1 ▢▢▢

Identify the correct equilibrium expression for the dissociation of barium nitrate.

A $\dfrac{[BaNO_3]}{[Ba^{2+}][NO_3^-]^2}$

B $[Ba^{2+}][NO_3^-]^2$

C $\dfrac{[Ba^{2+}][NO_3^-]^2}{[BaNO_3]}$

D $[Ba][NO_3]$

Question 2 ▢▢▢

A student completed a set of precipitation reactions and got the following results (ppt = precipitate).

Ion	Add 0.1 mol L^{-1} Na$_2$CO$_3$	Add 0.1 mol L^{-1} HCl	Add 0.1 mol L^{-1} KSCN	Add 0.1 mol L^{-1} AgNO$_3$
Ca^{2+}	White ppt	No change	No change	No change
Ba^{2+}	White ppt	No change	No change	No change
Pb^{2+}	White ppt	White ppt	No change	No change
Fe^{3+}	Brown ppt	No change	Red colour	No change
Cl$^-$	No change	No change	No change	White ppt

When testing an unknown solution, the student obtained the following results.

Add 0.1 mol L^{-1} Na$_2$CO$_3$	Add 0.1 mol L^{-1} HCl	Add 0.1 mol L^{-1} KSCN	Add 0.1 mol L^{-1} AgNO$_3$
White ppt	White ppt	No change	White ppt

What is the identity of the solution?

A Iron(III) chloride

B Lead(II) chloride

C Barium chloride

D Calcium chloride

Question 3 ⬤◻◻

The table below displays the solubility data for various carbonates at a specific temperature.

Which carbonate is the most soluble at this temperature?

Carbonate	K_{sp}
$BaCO_3$	5.1×10^{-9}
$CaCO_3$	2.8×10^{-9}
$MgCO_3$	3.5×10^{-8}
$ZnCO_3$	1.4×10^{-11}

A $BaCO_3$

B $CaCO_3$

C $MgCO_3$

D $ZnCO_3$

Question 4 ⬤◻◻

What is the solubility expression (K_{sp}) for zinc hydroxide?

A $K_{sp} = \dfrac{[Zn^{2+}][OH^-]}{[Zn(OH)_2]}$

B $K_{sp} = \dfrac{[Zn(OH)_2]}{[Zn^{2+}][OH^-]}$

C $K_{sp} = [Zn^{2+}][2OH^-]$

D $K_{sp} = [Zn^{2+}][OH^-]^2$

Question 5 ⬤⬤◻

The solubility of calcium carbonate is 1.9×10^{-3} g/100 mL at 75°C.

What is the K_{sp} of calcium carbonate at this temperature?

A 5.0×10^{-4}

B 4.1×10^{-6}

C 3.6×10^{-8}

D 6.8×10^{-9}

Question 6 ⬤⬤◻

Which of the following chemicals can be used to differentiate sodium carbonate and barium nitrate?

A Potassium nitrate

B Potassium sulfide

C Potassium sulfate

D Potassium chloride

Question 7 ⬤⬤◻

150 mL of 0.2 mol L^{-1} copper(II) sulfate solution is mixed with 150 mL of 0.3 mol L^{-1} silver nitrate solution.

Which row of the table identifies whether a precipitate (ppt) forms and the reason?

	Will a ppt form?	Reason
A	Yes	$Q > K_{sp}$
B	Yes	$Q < K_{sp}$
C	No	$Q > K_{sp}$
D	No	$Q < K_{sp}$

Question 8

What occurs when a solution of sodium hydroxide is added to a saturated solution of barium hydroxide?

A The solubility of barium hydroxide decreases.

B The concentration of barium ions increases.

C The concentration of hydroxide ions remains the same because the solution is saturated.

D There is no change.

Question 9

What is the molar solubility of silver carbonate in a solution of $0.153 \, \text{mol L}^{-1}$ calcium carbonate?

A 5.53×10^{-11}

B 7.44×10^{-6}

C 1.69×10^{-11}

D 6.26×10^{-6}

Question 10   ©NESA   2020 SII Q20

The graph shows the concentration of silver and chromate ions which can exist in a saturated solution of silver chromate.

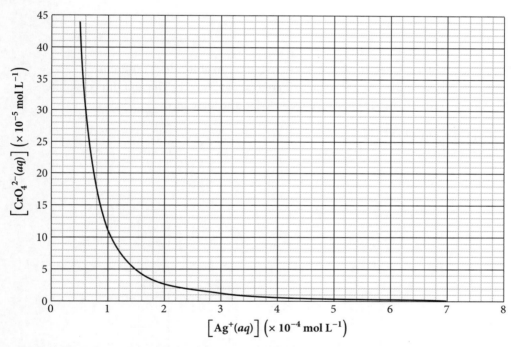

Based on the information provided, what is the K_{sp} for silver chromate?

A 1.1×10^{-8}

B 2.2×10^{-8}

C 1.1×10^{-12}

D 4.4×10^{-12}

Section II: Short-answer questions

> **Instructions to students**
> · Answer all questions in the spaces provided.

Question 11 (10 marks) ◖●●◗

Magnesium chloride can be dissolved in water.

a ◖●●◗ Describe the changes that occur in both bonding and entropy when magnesium chloride is dissolved in water. Support your answer with a labelled diagram.   4 marks

_____

_____

_____

_____

_____

_____

_____

_____

b ◖●●◗ A student investigated the solubility of magnesium chloride at different temperatures. The volume of water remained constant at 100 mL. The amounts of solute needed to make saturated solutions were recorded in the table below.

Temperature of water (°C)	10	20	35	50	75
Mass of solute (g)	53.6	54.3	58.8	63.5	68.0

Draw a graph of solubility versus temperature on the grid below.   3 marks

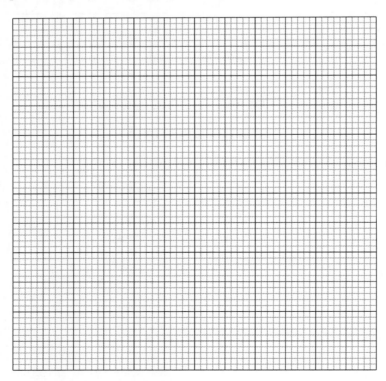

c [●○○] What mass of magnesium chloride will dissolve in 100 mL of water at 25°C?   1 mark

_____

d [●●○] Describe and explain what a student would observe if 125 g of magnesium chloride was mixed in a beaker containing 200 mL of water at a temperature of 40°C.   2 marks

_____

_____

_____

_____

Question 12 (5 marks)

There has been a mishap in the chemical storeroom and a dropper bottle has lost its label. Based on the bottle's location, the solution could be sodium hydroxide, barium nitrate or sodium chloride.

To help identify the solution, a student wrote out the following table (ppt = precipitate).

Chemical to add	Possible solutions		
	Sodium hydroxide (aq)	Barium nitrate (aq)	Sodium chloride (aq)
Silver nitrate (aq)	Ppt	No ppt	Ppt
Sodium hydroxide (aq)	No ppt	Ppt	No ppt
Copper sulfate (aq)	Ppt	Ppt	No ppt

a [●●○] Construct a flow chart showing a method the student could follow to identify the solution.   3 marks

b [●○○] Write a balanced net ionic equation for the reaction that would identify barium nitrate.   2 marks

_____

_____

Question 13 (9 marks)

The K_{sp} of lead(II) chloride is 2.4×10^{-4} at 25°C.

a ●● Calculate the maximum concentration of lead ions in a solution that contains $0.025 \, \text{mol L}^{-1}$ of Cl^-.　　2 marks

b ●●● Will a precipitate of lead(II) chloride form when $100 \, \text{mL}$ of $3.0 \times 10^{-2} \, \text{mol L}^{-1}$ lead(II) nitrate is added to $400 \, \text{mL}$ of $9.0 \times 10^{-2} \, \text{mol L}^{-1}$ of sodium chloride? Justify your answer.　　4 marks

c ●●● The solubility of lead(II) chloride increases to $1.94 \, \text{g}/100 \, \text{mL}$ at 60°C.

Determine the K_{sp} at this temperature.　　3 marks

CHAPTER 2
MODULE 6: ACID/BASE REACTIONS

Test 4: Properties of acids and bases

Section I: 10 marks. Section II: 20 marks. Total marks: 30.
Suggested time: 54 minutes

Section I: Multiple-choice questions

Instructions to students
- For each question, circle the multiple-choice letter to indicate your answer.

Question 1

Which list contains only basic substances?

A Baking soda, urine, ammonia

B Drain cleaner, facial cleanser, orange juice

C Antacid, dishwashing detergent, sea water

D Lemonade, bleach, tomato juice

Question 2

Which of the following names one monoprotic acid and one diprotic acid?

A Hydrochloric acid and acetic acid

B Nitric acid and sulfuric acid

C Sulfuric acid and phosphoric acid

D Carbonic acid and citric acid

Question 3

Indicators

A are oxidising agents.

B change colour at specific pH values.

C consist of strong acids or bases.

D consist of salt solutions.

Question 4 ◑◌◌

An indicator is red in acid, green in base and yellow in neutral solutions. This indicator was added to a sample of sodium hydroxide. Nitric acid was slowly added until it was in excess.

Which of the following shows the colour changes of the solution?

A Stays green

B Yellow to red to green

C Green to yellow to red

D Green to yellow

Question 5 ◑◑◌

The diagram shows the pH values of some substances.

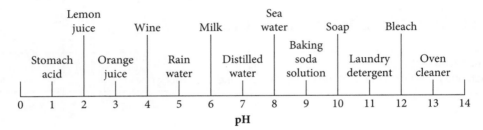

The concentration of hydroxide ions in

A orange juice is 100 times higher than in rain water.

B laundry detergent is 10 times lower than in soap.

C milk is the same as in distilled water.

D sea water is 1000 times higher than in rain water.

Question 6 ◑◑◌

The table shows the pH range and colour changes of three acid–base indicators.

Indicator	pH	Colour	pH	Colour
Methyl orange	3.2	Red	4.4	Yellow
Bromothymol blue	6.0	Yellow	7.6	Blue
Phenolphthalein	8.2	Colourless	10.0	Pink

Which alternative correctly identifies the colour of the solution at pH 7.3?

	Methyl orange	**Bromothymol blue**	**Phenolphthalein**
A	Yellow	Green	Colourless
B	Yellow	Blue	Colourless
C	Red	Green	Colourless
D	Red	Blue	Colourless

Question 7 ⬤⬤◯

Which of the following equations correctly corresponds to the acid–base theory of the scientists who proposed it?

	Equation	Theory
A	$NH_3(aq) + CH_3OH \rightleftharpoons CH_3O^-(aq) + NH_4^+(aq)$	Brønsted–Lowry
B	$H^+(aq) + OH^-(aq) \rightarrow H_2O(l)$	Brønsted–Lowry
C	$HNO_3(aq) + H_2O(l) \rightleftharpoons H_3O^+(aq) + NO_3^-(aq)$	Arrhenius
D	$Na(s) + HCl(g) \rightarrow NaCl(s) + H_2(g)$	Arrhenius

Question 8 ©NESA 2018 SI Q9 ⬤⬤◯

Which of the following would **not** have been classified as an acid by Antoine Lavoisier in 1780?

A Acetic acid

B Citric acid

C Sulfuric acid

D Hydrochloric acid

Question 9 ⬤⬤⬤

100 mL of a $0.5\,mol\,L^{-1}$ solution of sodium hydroxide was mixed with an equal volume of a $0.5\,mol\,L^{-1}$ solution of hydrochloric acid. Before mixing, both solutions were at the same temperature. After mixing, the temperature increased by 3.4°C.

What is the molar heat of neutralisation?

A $-28.42\,kJ\,mol^{-1}$

B $-56.85\,kJ\,mol^{-1}$

C $-142.12\,kJ\,mol^{-1}$

D $-158.6\,kJ\,mol^{-1}$

Question 10 ⬤⬤⬤

What mass of sodium hydrogen carbonate is required to neutralise $200\,mL$ of $0.4\,mol\,L^{-1}$ nitric acid?

A 4.24 g

B 6.72 g

C 8.48 g

D 13.44 g

Section II: Short-answer questions

Instructions to students
- Answer all questions in the spaces provided.

Question 11 (4 marks)

The following table provides some information about three natural indicators.

Indicator	pH range	Acid colour	Base colour
Turmeric	7.4–8.6	Yellow	Red
Blueberries	2.8–3.2	Red	Blue
Red roses	6.5–7.2	Pink	Green

a Outline a method you would use to make one of these natural indicators.   2 marks

A student used an indicator made from turmeric on the following household substances and used their results to classify them as acidic or basic.

Substance	Colour with indicator	Classification
Hydrochloric acid	Yellow	Acid
Vinegar	Yellow	Acid
Water	Yellow	Acid
Baking soda	Yellow	Acid
Sodium hydroxide	Red	Base

b Discuss the validity of the student's results.   2 marks

Question 12 (8 marks)

The diagram shows a coffee cup calorimeter used by a student to measure the enthalpy of neutralisation in an acid–base reaction.

The student combined 65 mL of 0.20 mol L^{-1} sodium hydroxide and 130 mL of 0.10 mol L^{-1} sulfuric acid. Both solutions were initially at a temperature of 19.2°C.

After mixing, the final temperature was 24.3°C.

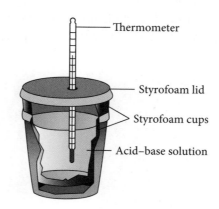

Thermometer

Styrofoam lid

Styrofoam cups

Acid–base solution

a Calculate the molar heat of neutralisation for this reaction.   4 marks

_____

_____

_____

_____

_____

_____

_____

_____

_____

_____

_____

b Identify the limiting reactant and number of unreacted moles.   2 marks

_____

_____

_____

c List **two** assumptions made by the student while performing these calculations.   2 marks

_____

_____

Question 13 (3 marks) ⬤⬤⬤

Explain why hydrochloric acid dissolved in methanol is defined as a Brønsted–Lowry acid but cannot be defined as an Arrhenius acid. Support your answer with a chemical equation.

_____

_____

_____

_____

_____

Question 14 (5 marks) ⬤⬤⬤

The amount of carbon dioxide in Earth's atmosphere is increasing, leading to more carbon dioxide dissolving in the oceans.

a   Explain why increasing levels of carbon dioxide are making oceans more acidic. Include balanced equations in your answer.   4 marks

_____

_____

_____

_____

_____

_____

b   As ocean waters become more acidic, animals with shells (which are made from calcium carbonate) are becoming more vulnerable to predators.

State **one** reason for this.   1 mark

_____

_____

_____

Test 5:   Using Brønsted–Lowry theory

> Section I: 10 marks. Section II: 25 marks. Total marks: 35.
> Suggested time: 63 minutes

Section I: Multiple-choice questions

> **Instructions to students**
> - For each question, circle the multiple-choice letter to indicate your answer.

Question 1

For a sample of pure water, which statement is always correct regardless of conditions?

A $[H^+] = [OH^-]$

B pH depends on temperature.

C $[OH^-] = 1.0 \times 10^{-7} \, mol \, L^{-1}$

D $[H_3O^+] = 1.0 \times 10^7 \, mol \, L^{-1}$

Question 2

Strong acids

A are always at high concentrations.

B have large K_a values.

C have large pK_a values.

D do not completely ionise in solution.

Question 3

Which of the following pairs includes a strong acid and a weak acid?

A Hydrochloric acid and sodium hydroxide

B Sulfuric acid and ammonia

C Nitric acid and acetic acid

D Phosphoric acid and formic acid

Question 4

Based on the following information, which of the following acids is stronger and which has the higher pH?

Carbonic acid, $K_a = 4.4 \times 10^{-7}$

Ethanoic acid, $K_a = 1.8 \times 10^{-5}$

	Stronger acid	Higher pH
A	Carbonic acid	Carbonic acid
B	Carbonic acid	Ethanoic acid
C	Ethanoic acid	Carbonic acid
D	Ethanoic acid	Ethanoic acid

Question 5 ●●●

Which of the following acids will ionise to produce the strongest conjugate base?

A HCN

B HBr

C H_2SO_4

D H_3PO_4

Question 6 ●●●

What change occurs when an acid is diluted with water?

A Ionisation increases and pH decreases.

B Ionisation increases and pH increases.

C Ionisation decreases and pH decreases.

D Ionisation decreases and pH increases.

Question 7 ●○○

What is the pH of a $0.021\,\text{mol}\,L^{-1}$ solution of hydrochloric acid?

A 0.95

B 1.00

C 1.60

D 1.70

Question 8 ●○○

What is the concentration $(\text{mol}\,L^{-1})$ of hydroxide ions in a solution that has a pH of 7.4?

A 3.98×10^{-8}

B 2.51×10^{-7}

C 8.69×10^{-1}

D 8.19×10^{-1}

Question 9 ●●○

What is the pH of a $0.083\,\text{mol}\,L^{-1}$ solution of sodium hydroxide?

A 1.08

B 1.20

C 12.90

D 12.80

Question 10 ©NESA 2018 SI Q17 ●●●

Increasing amounts of carbon dioxide were dissolved in two beakers, one containing water and one a mixture of water and a buffer. The pH in each beaker was measured and the results were graphed.

Which graph best represents the results?

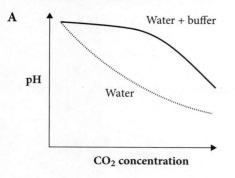

A

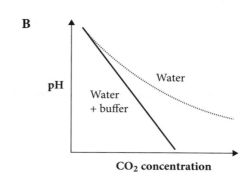

B

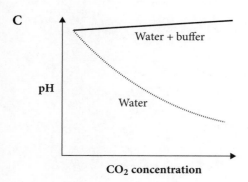

C

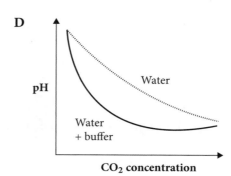

D

Section II: Short-answer questions

> **Instructions to students**
> • Answer all questions in the spaces provided.

Question 11 (3 marks) ●●●

Draw a labelled diagram that suitably models the differences between strong, weak, dilute and concentrated acids.

Question 12 (6 marks)

a ◖◗ A volumetric flask contains 100 mL of 0.230 mol L^{-1} potassium hydroxide solution. A 20 mL aliquot is poured into a second beaker.

Determine [OH$^-$] in this sample.   1 mark

b ◖◗◗ The 20 mL aliquot in part **a** is diluted to 100 mL. Determine the pH of this sample.   3 marks

c ◖◗◗ Explain the effect dilution has on the pH of a basic solution.   2 marks

Question 13 (7 marks)

A chemist has prepared one beaker containing 50.0 mL of a 0.010 mol L^{-1} solution of nitric acid and a second beaker containing 50.0 mL of a 0.010 mol L^{-1} solution of nitrous acid.

a ◖◗ Determine the pH of the nitric acid solution.   1 mark

b ◖◗◗ Although the nitrous acid solution is the same volume and concentration as the nitric acid solution, the pH values of these acids are different.

Why is this the case?   2 marks

c   The chemist added 50 mL of a 0.010 mol L^{-1} solution of calcium hydroxide to the beaker containing nitric acid.

Determine the pH of the final solution.   4 marks

_____

_____

_____

_____

_____

_____

_____

_____

_____

_____

Question 14 (5 marks)

Hydrogen sulfate ions (HSO_4^-) are classified as amphiprotic.

a   Define 'amphiprotic'; include chemical equations in your answer.   3 marks

_____

_____

_____

_____

_____

_____

_____

b   Outline how a student could confirm the amphiprotic nature of the hydrogen sulfate ion.   2 marks

_____

_____

_____

_____

_____

Question 15 (4 marks) ●●●

A student measured the pH values of sodium chloride (NaCl), ammonium nitrate (NH_4NO_3) and sodium ethanoate (CH_3COONa).

Because of careless recording, the student did not write the name of each solution next to its recorded pH.

Complete the student's results table, using chemical equations to justify your choice.

Solution	pH	Justification
	6.9	
	8.6	
	5.4	

Test 6: Quantitative analysis

Section I: 10 marks. Section II: 25 marks. Total marks: 35.
Suggested time: 63 minutes

Section I: Multiple-choice questions

Instructions to students
- For each question, circle the multiple-choice letter to indicate your answer.

Question 1

Which piece of glassware is **not** required in the preparation of a primary standard solution?

A Beaker

B Measuring cylinder

C Volumetric flask

D Filter funnel

Question 2

A student performed an acid–base titration. They placed the acid in a burette and used a pipette to transfer 25 mL of base into a conical flask.

What liquids should the student use in the cleaning procedure for these three pieces of glassware?

	Burette	**Pipette**	**Conical flask**
A	Distilled water	Base	Distilled water
B	Distilled water	Distilled water	Base
C	Acid	Base	Distilled water
D	Acid	Distilled water	Base

Use the following information to answer Questions 3 and 4.

Tartaric acid is the predominant acid in wine. It is a diprotic acid with a molar mass of $150.087\,\text{g mol}^{-1}$. Before wine can be titrated, deactivated carbon is added to remove the colour and then the wine is passed through a strong vacuum to degas the wine, removing any carbon dioxide.

Question 3

A student titrated 25.0 mL samples of wine with $0.200\,\text{mol L}^{-1}$ sodium hydroxide solution. The average titre was 31.4 mL.

What was the concentration of tartaric acid in the wine?

A Greater than $0.200\,\text{mol L}^{-1}$

B Equal to $0.200\,\text{mol L}^{-1}$

C Less than $0.200\,\text{mol L}^{-1}$

D Without the formula for tartaric acid, the concentration cannot be calculated.

Question 4 ⬤⬤▨

The end point of this titration occurred at a pH of 8.2.

Which indicator is the most suitable?

A Phenolphthalein

B Bromothymol blue

C Methyl orange

D Methyl red

Question 5 ⬤⬤▨

What is the pH at the equivalence point in the following titration curve?

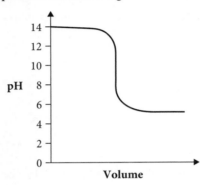

A 14

B 13

C 12

D 10

Question 6 ⬤⬤▨

The pK_b of ammonia is 4.75. What is the dissociation constant (K_b)?

A 6.77×10^{-1}

B 1.78×10^{-5}

C 5.62×10^{-10}

D 9.25

Question 7 ⬤⬤▨

A buffer was made from equal volumes of a $0.1 \, mol \, L^{-1}$ HNO_3 solution and a $0.1 \, mol \, L^{-1}$ $NaNO_3$ solution.

Which row of the table correctly evaluates the effectiveness of this buffer?

	Effectiveness	Justification
A	Ineffective	HNO_3 is a strong acid
B	Ineffective	$NaNO_3$ forms a solution with a pH of 7
C	Effective	$NaNO_3$ forms a solution with a pH of 7
D	Effective	NO_3^- is the conjugate base of HNO_3

Question 8 ●●

Which of the following conductivity graphs represents the titration of sodium hydroxide with sulfuric acid?

A

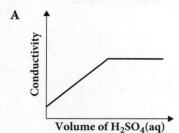

B

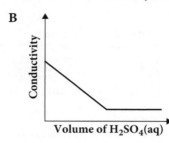

C

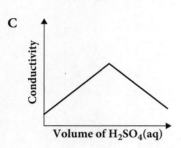

D

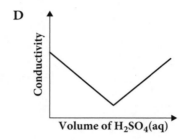

Question 9 ●●●

Which of the following titration curves represents a volume of hydrochloric acid added to a buffer solution?

A

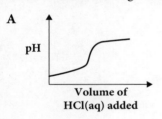

B

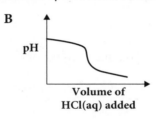

C

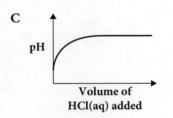

D

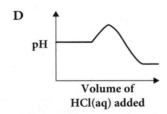

Question 10 ●●●

A buffer solution contains lactic acid and its conjugate base. The pK_a of lactic acid is 3.86.

What is the pH at which the solution buffers?

A 2.0

B 3.0

C 4.0

D 5.0

Section II: Short-answer questions

Instructions to students
· Answer all questions in the spaces provided.

Question 11 (6 marks)

A titration is carried out using sodium hydroxide to determine the concentration of a solution of sulfuric acid.

a ◐◌◌ Evaluate the use of sodium hydroxide as a primary standard.   3 marks

_____

_____

_____

_____

_____

_____

_____

b ◐◌◌ Sketch the shape you might expect of the pH curve for this titration.   1 mark

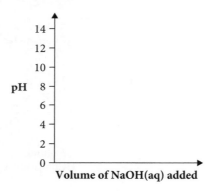

c ◐◐◌ Based on your titration curve, identify a suitable indicator that could be used by a student in a school laboratory. Justify your choice.   2 marks

_____

_____

_____

_____

Question 12 (6 marks) ●●●

Apple cider vinegar is made by the fermentation of apple juice and contains acetic acid.
The label of a popular apple cider vinegar is shown below.

To test this claim, a student diluted 20 mL of apple cider vinegar to 200 mL.

25 mL of the diluted solution was titrated with a 0.0145 mol L^{-1} solution of sodium hydroxide, using a phenolphthalein indicator. An average titre of 22.5 mL was required.

Using these values, determine whether the manufacturer's claim is correct.

Question 13 (7 marks) ⬤⬤⬤

In a conductometric titration, ammonia was added to 200 mL of a 1.12×10^{-4} mol L^{-1} solution of sulfuric acid. The results of the titration are shown below.

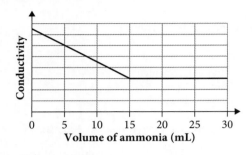

a Explain the shape of this graph.   3 marks

_____

_____

_____

_____

_____

b Calculate the concentration of ammonia.   4 marks

_____

_____

_____

_____

_____

_____

_____

_____

Question 14 (6 marks) ⬤⬤⬤

A student made a buffer solution by mixing equal volumes of equimolar citric acid ($C_6H_8O_7$) and sodium citrate ($Na_3C_6H_5O_7$). They then added a few drops of universal indicator. When small volumes of a strong acid or a strong base were added, no colour change was observed.

a Using relevant chemical equations, explain this observation.   4 marks

_____

_____

_____

_____

_____

_____

b Increasing the concentration of this buffer solution increases its buffering capacity. Explain this statement.   2 marks

_____

_____

_____

_____

_____

CHAPTER 3
MODULE 7: ORGANIC CHEMISTRY

Test 7: Hydrocarbons and products of reactions involving hydrocarbons

Section I: 10 marks. Section II: 25 marks. Total marks: 35.
Suggested time: 63 minutes

Section I: Multiple-choice questions

Instructions to students
· For each question, circle the multiple-choice letter to indicate your answer.

Question 1 ⬤◗▮

The structure of a compound is shown.

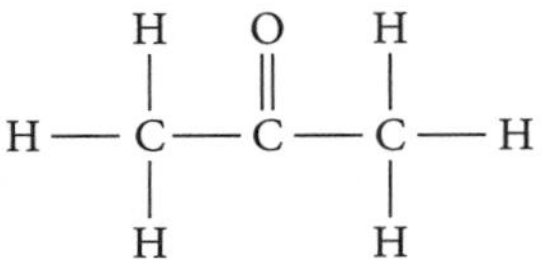

What is the IUPAC name of this compound?

A Acetone

B Propanal

C Propanone

D Propan-2-one

Use the following diagram to answer Questions 2 and 3.

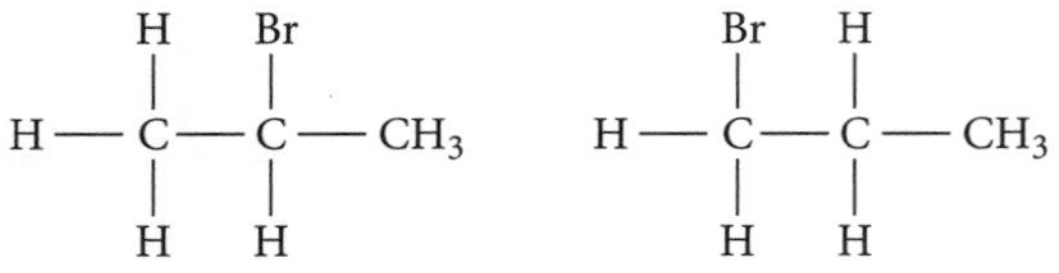

Question 2 ⬤◗▮

X and Y are best described as

A isotopes.

B chain isomers.

C functional group isomers.

D positional isomers.

Question 3 ⚫⚫⚫

Which alternative correctly identifies the reactants used to make X and Y?

	X	Y
A	Prop-2-ene + Br_2	Prop-1-ene + Br_2
B	Prop-2-ene + HBr	Prop-1-ene + HBr
C	Propene + HBr	Propene + HBr
D	Propene + Br_2	Propene + Br_2

Question 4 ⚫⚪⚪

The structure of a compound is shown.

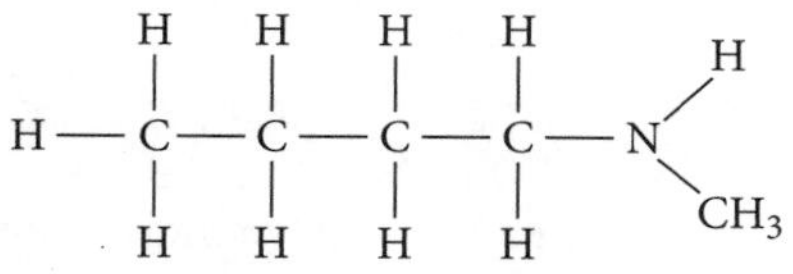

What is the IUPAC name of this compound?

A *N*-Methylbutanamide

B *N*-Methylbutanamine

C *N*-Methanylamide

D *N*-Butylmethanamine

Question 5 ⚫⚫⚫

A gaseous unsaturated hydrocarbon with a molar mass of $28.052\,\text{g mol}^{-1}$ was bubbled through dilute sulfuric acid. The mixture was distilled to obtain a pure liquid, Y, with a molar mass of $46.068\,\text{g mol}^{-1}$, which was soluble in water.

What is the identity of the functional group of the pure liquid Y?

A Alcohol

B Aldehyde

C Amine

D Carboxylic acid

Question 6 ⚫⚫⚪

Which alternative correctly identifies the geometry around each numbered carbon atom in the compound shown?

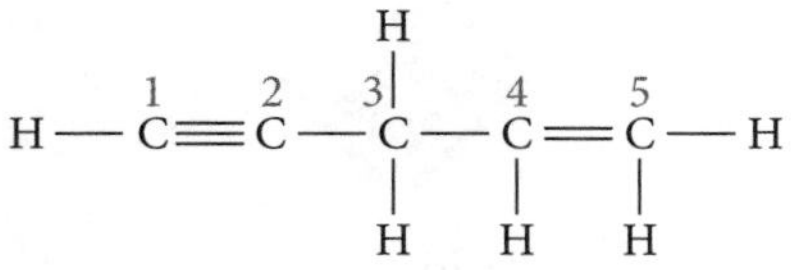

	C1	C2	C3	C4	C5
A	Trigonal planar	Trigonal planar	Tetrahedral	Linear	Linear
B	Linear	Linear	Tetrahedral	Trigonal planar	Trigonal planar
C	Linear	Linear	Trigonal planar	Trigonal planar	Trigonal planar
D	Trigonal planar	Trigonal planar	Trigonal planar	Linear	Linear

Question 7 ●●●

The graph shows the boiling points of straight chain and branched chain alkanes.

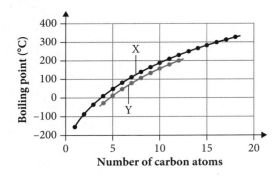

Which alternative best describes the graph?

	X chain	Y chain	Explanation
A	Branched	Straight	Branched chain alkanes have higher boiling points because these molecules can pack closer together and allow many dispersion forces to operate.
			They require more energy than straight chain molecules that cannot pack closely together.
B	Branched	Straight	Branched chain alkanes have higher boiling points because these molecules have larger molar mass because of the side branches than straight chain molecules.
			This means they have more dispersion forces.
C	Straight	Branched	Straight chain alkanes have higher boiling points because these molecules can pack closer together and allow many dispersion forces to operate.
			They require more energy than branched chain molecules that cannot pack closely together.
D	Straight	Branched	Straight chain alkanes have higher boiling points because these molecules have lower molar mass and can pack closer together.
			They allow many dispersion forces compared with branched chain molecules, which have larger molar masses.

Question 8 ●●●

A student had some cyclohexane left over at the end of an experiment. A section of the safety data sheet (SDS) for cyclohexane is shown with its Globally Harmonised System (GHS) labels.

The GHS guides users in the handling and disposal of substances.

Flammable

Health hazard

Irritant

Which alternative correctly identifies how the student should handle and manage the excess cyclohexane?

A It should be handled in a fume cupboard and poured back into the original reagent bottle.

B It should be handled next to open windows and poured down the sink.

C It should be handled in a fume cupboard and poured into a waste organic solvent bottle.

D It should be handled in outdoor areas and neutralised with some baking powder before pouring it down the sink.

Question 9 ⬤⬤◯

Which alternative correctly describes possible ways in which an organic compound may enter the body?

    I Inhalation to the lungs

    II Absorption through the skin

    III Ingestion

A I and II

B I and III

C II and III

D I, II and III

Question 10 ⬤⬤⬤

At 25°C and 100 kPa, 5.726 L of hydrogen gas was reacted with 6.473 g of an alkene in the presence of a nickel catalyst.

What is the mass of the product formed?

A 6.946 g

B 7.577 g

C 28.02 g

D 30.068

Section II: Short-answer questions

> **Instructions to students**
> - Answer all questions in the spaces provided.

Question 11 (6 marks) ⬤⬤◯

The table lists some properties of methane and petrol.

Property	Methane	Petrol
Average molar mass	16	114
ΔH (kJ mol^{-1})	74.85	5460
Density (g mL^{-1})	0.465	0.690

a A car has a 60.0 L fuel tank. Calculate the energy released by the combustion of a full tank of petrol.
    3 marks

b Calculate the volume of methane gas at 25°C and 100 kPa required to supply the same amount of energy as a 60.0 L tank of petrol.   3 marks

_____

_____

_____

_____

_____

_____

Question 12 (4 marks)

State the IUPAC names of C_4H_8 and the compounds labelled A to C in the flow chart and draw the structural formulae. (Note: Only one product forms in each step.)

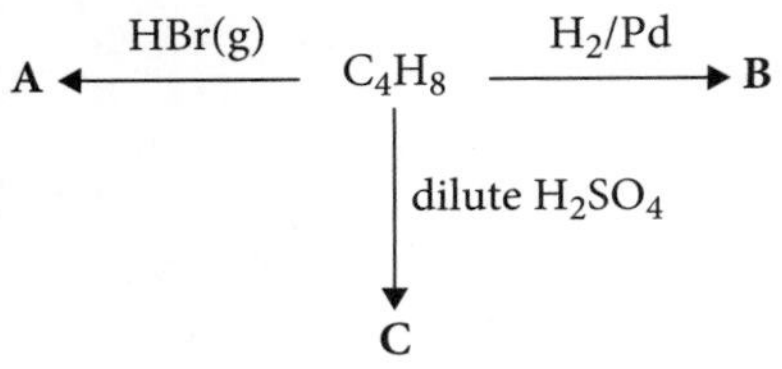

Compound	IUPAC name	Structural formula
C_4H_8		
A		
B		
C		

Question 13 (4 marks) ●●●

An unsaturated hydrocarbon has a molar mass $56.104 \, \text{g mol}^{-1}$. Draw structural formulae for the hydrocarbon and its isomers. Write IUPAC names for all structures.

Question 14 (4 marks)

Poor management in handling of chlorinated solvents in the past has resulted in a number of contaminated sites in New South Wales. Trichloroethylene is one of the most frequently detected chlorinated solvents in ground water.

a ●●● Draw the structural formula of trichloroethylene.   1 mark

b ●● The diagram shows how chlorinated solvents are transferred from transport trucks to storage tanks.

Some properties of chlorinated solvents are that they are:

- denser than water
- immiscible with water
- slow to degrade and toxic
- able to seep through concrete.

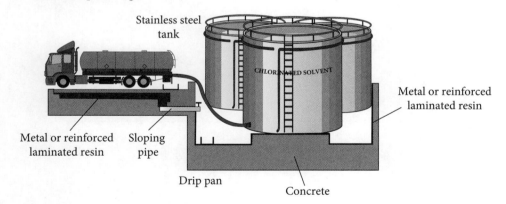

Analyse the physical structure and positions of the truck and the storage tank in relation to the safe transfer and storage of chlorinated solvents.   3 marks

_____

_____

_____

_____

_____

Question 15 (3 marks)

The labels have fallen off two reagent bottles containing two different colourless liquids. The labels were hex-3-ene and hex-1-ene. Outline a chemical procedure for identifying the contents of each bottle.

_____

_____

_____

_____

_____

_____

Question 16 (4 marks)

Draw the structural formulae and state the IUPAC names of the positional isomers of a straight chain alcohol with five carbons.

Name	Drawing

Test 8:   Alcohols

Section I: 10 marks. Section II: 25 marks. Total marks: 35.
Suggested time: 60 minutes

Section I: Multiple-choice questions

Instructions to students
- For each question, circle the multiple-choice letter to indicate your answer.

Question 1

What is the IUPAC name of the compound shown?

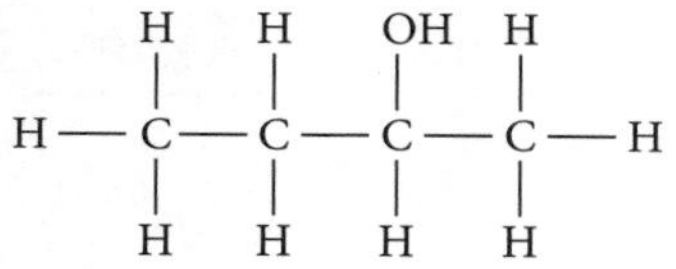

A Butanol

B Butan-2-ol

C Pentan-2-ol

D Propanol

Question 2

What is the IUPAC name of the compound shown?

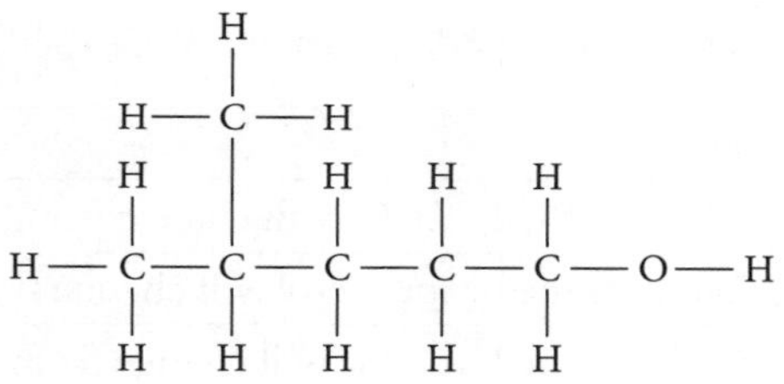

A Hexan-1-ol

B 2-Methylpentan-5-ol

C 4-Methylpentan-1-ol

D 4-Ethylpentan-1-ol

Question 3

When an alkanol was reacted with concentrated sulfuric acid, a mixture of two alkenes was produced.
What is the identity of the alkanol?

A Ethanol

B Propan-1-ol

C Propan-2-ol

D Butan-2-ol

Use the following information to answer Questions 4 and 5.

Propan-1-ol was reacted with Y and $K_2Cr_2O_7$ to produce X.

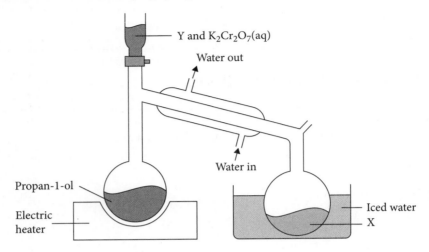

Question 4 ⬤⬤▨

What are the identities of X and Y?

	X	Y
A	Propanal	Dilute H_2SO_4
B	Propanal	Concentrated H_2SO_4
C	Propanone	Dilute H_2SO_4
D	Propanone	Concentrated H_2SO_4

Question 5 ⬤⬤⬤

Which alternative correctly identifies the independent variable and experimental observation?

	Independent variable	Observation
A	$K_2Cr_2O_7$	The colour of the $K_2Cr_2O_7$ will change from orange to green.
B	$K_2Cr_2O_7$	The colour of the propan-1-ol will change from orange to green.
C	Propan-1-ol	The colour of the $K_2Cr_2O_7$ will change from orange to green.
D	Propan-1-ol	The colour of the propan-1-ol will change from orange to green.

Question 6 ⬤⬤▨

A chemist reacts a substance, X, which is soluble in water, with a gaseous substance, Y, to produce bromoethane and water. Which alternative correctly identifies X and Y, and the type of reaction?

	X	Y	Type of reaction
A	Ethene	Br_2	Substitution
B	Ethene	HBr	Addition
C	Ethanol	Br_2	Addition
D	Ethanol	HBr	Substitution

Question 7 ⬤⬤⬜

The graph shows the mass of carbon dioxide released during fermentation. What is the volume of ethanol produced, given its density is $0.79\,g\,mL^{-1}$?

A 7.4 mL

B 9 mL

C 13 mL

D 19 mL

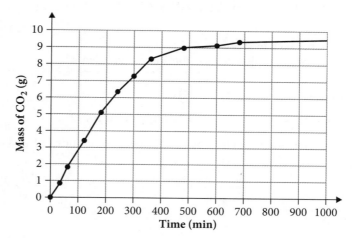

Question 8 ⬤⬤⬤

Some bond energies are given in the table.

Which of the following compounds will react the fastest?

A Ethane

B Bromoethane

C Chloroethane

D Iodoethane

Bond	Bond energy $(kJ\,mol^{-1})$
C–H	412
C–C	348
C–Br	338
C–I	238

Question 9 ⬤⬜⬜

What is the name of the process shown?

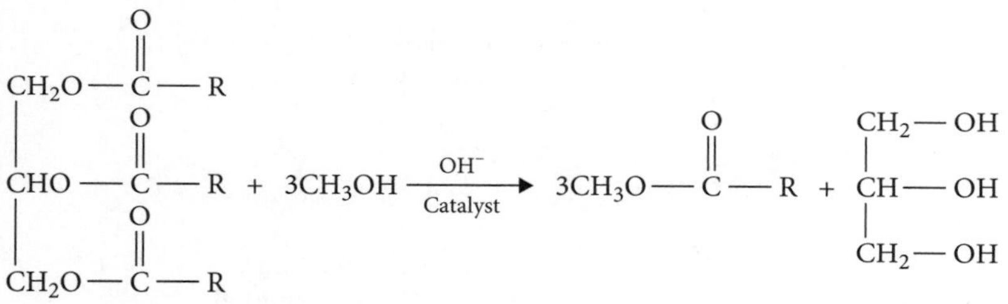

$$CH_2O-\overset{\overset{\displaystyle O}{\|}}{C}-R$$
$$CHO-\overset{\overset{\displaystyle O}{\|}}{C}-R \;+\; 3CH_3OH \xrightarrow[\text{Catalyst}]{OH^-} 3CH_3O-\overset{\overset{\displaystyle O}{\|}}{C}-R \;+\; \begin{matrix}CH_2-OH\\ \mid \\ CH-OH\\ \mid \\ CH_2-OH\end{matrix}$$
$$CH_2O-\overset{\overset{\displaystyle O}{\|}}{C}-R$$

A Condensation

B Esterification

C Saponification

D Transesterification

Question 10 ⬤⬤⬜

A chemist added an unlabelled alcohol to acidified potassium permanganate and refluxed the mixture. There was no colour change. What is the identity of the alcohol?

A Ethanol

B Butan-2-ol

C 2-Methylpropan-2-ol

D 2-Methylpentan-1-ol

Section II: Short-answer questions

Instructions to students
· Answer all questions in the spaces provided.

Question 11 (3 marks) ⬤⬤⬤

Ethanol can be produced in a number of ways.

Complete the flow chart by identifying compounds 1–4 and reagents 1–4.

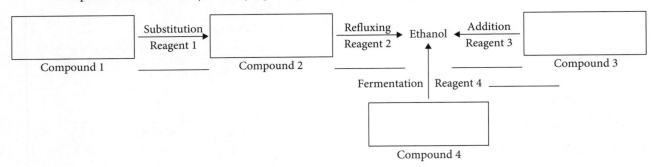

Question 12 (13 marks)

A student set up an experiment to compare the molar heat of combustion of octane and ethanol. The student's experimental set-up is shown below.

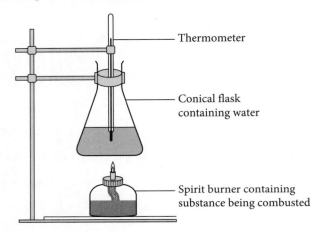

a ⬤⬤⬤ Every time the student repeated the experiment, they found more soot at the bottom of the conical flask when octane was combusted and less soot for the combustion of ethanol.

Explain this observation. Include relevant equations in your response.　　3 marks

_____

_____

_____

_____

_____

_____

b 🔲 The student placed some ethanol in the spirit burner and some water in the conical flask. They lit the spirit burner and recorded the temperature of the water every 30 seconds for 4 minutes, as shown.

Time (min)	0	0.5	1	1.5	2	2.5	3	3.5	4
Temp. (°C)	22.0	24.2	26.3	28.5	30.5	32.0	30.5	29.4	28.0

Graph the data on the grid provided.   3 marks

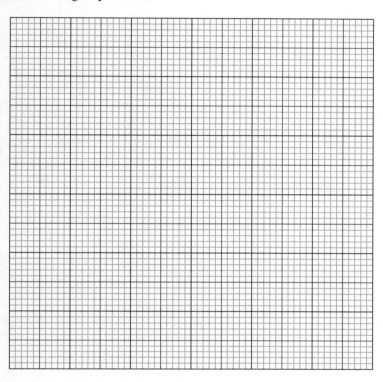

c 🔲 After how many minutes is it likely that the student extinguished the spirit burner? Explain your response.   2 marks

_____

_____

_____

_____

d 🔲 The student also recorded the following data:

Initial mass of spirit burner = 252.28 g
Final mass of spirit burner = 251.86 g
ΔH ethanol = 745 kJ mol^{-1}

Assuming the density of water is 0.996 g mL^{-1}, determine the volume of the water in the conical flask, using relevant data provided here and in part **b**.   3 marks

_____

_____

_____

_____

_____

e   [○ ◼] The theoretical value for the molar heat of combustion of ethanol is $1367\,\text{kJ}\,\text{mol}^{-1}$.
Comment on the validity of the data and suggest improvements to the experimental set-up.   2 marks

_____

_____

_____

_____

_____

_____

Question 13 (6 marks) [○ ○ ◼]

The labels have fallen off three reagent bottles. The labels are for propan-1-ol, propan-2-ol
and 2-methylpropan-2-ol.

a   Outline a procedure for identifying the contents of each bottle.   3 marks

_____

_____

_____

_____

_____

_____

_____

_____

b   Write the half-equations and a net equation for the oxidation of propan-2-ol by acidified
potassium permanganate.   3 marks

Reduction: _____

Oxidation: _____

Net: _____

Question 14 (3 marks) [○ ◼]

Outline the production of biofuels and comment on their viability as fuels in the future
in Australia.

_____

_____

_____

_____

_____

Test 9: Reactions of organic acids and bases

Section I: 10 marks. Section II: 25 marks. Total marks: 35.
Suggested time: 60 minutes

Section I: Multiple-choice questions

Instructions to students
- For each question, circle the multiple-choice letter to indicate your answer.

Question 1

What is the IUPAC name of the compound shown?

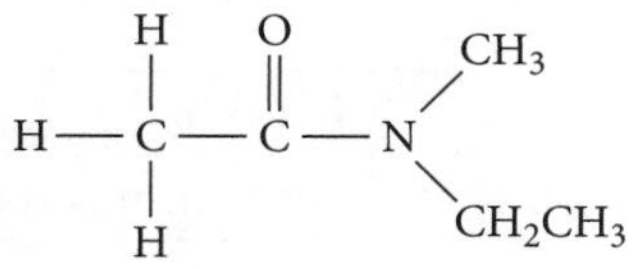

A N,N-Diethylmethylethanamide

B N-Ethyl-N-methylethanamide

C N-Methyldiethylethanamide

D N,N-Dimethylethanamide

Question 2

Which one of these substances would turn litmus blue?

A Methanol

B Methanamide

C Methanamine

D Methanoic acid

Use the following diagram to answer Questions 3 and 4.

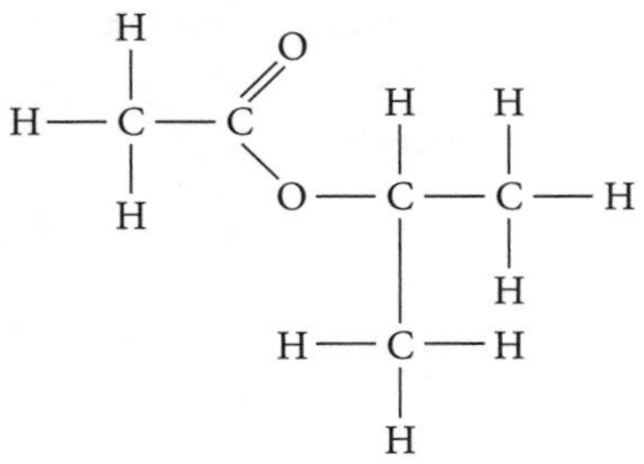

Question 3

Which alternative correctly lists the chemicals required to make the compound shown?

A Ethanol, propanoic acid, dilute sulfuric acid

B Ethanol, propan-2-ol, concentrated sulfuric acid

C Ethanoic acid, propan-1-ol, concentrated sulfuric acid

D Ethanoic acid, propan-2-ol, concentrated sulfuric acid

Question 4 ●●●

The equipment shown can be used to separate the compound in the diagram from the reaction mixture.

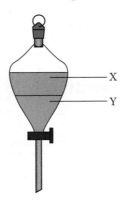

Which alternative correctly identifies the name of the equipment and the properties of layers X and Y?

	Name of equipment	Layer X	Layer Y
A	Funnel	Turns litmus blue	Turns litmus red
B	Standing funnel	No effect on litmus	Turns litmus blue
C	Separating funnel	No effect on litmus	Turns litmus red
D	Separating flask	Turns litmus blue	Turns litmus red

Question 5 ●●○

The graph shows boiling points for alkanes, alcohols, carboxylic acids and ketones of a number of carbon chain lengths.

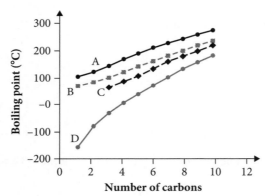

Which line represents the boiling points of alcohols?

A Line A

B Line B

C Line C

D Line D

Question 6

The diagram shows a structure that forms when grease is removed from a surface. Grease is in the middle of the structure.

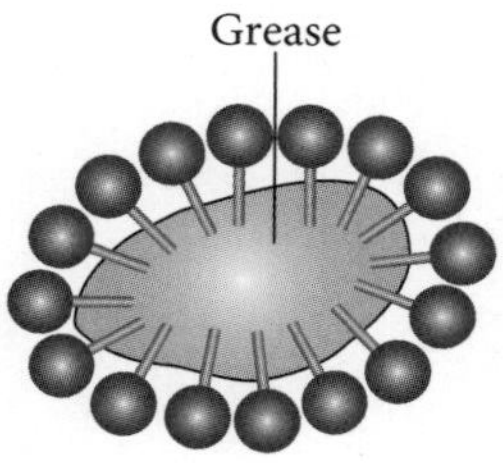

Which statement is correct about the structure shown? It is

A an emulsion.

B a micelle.

C a soap ion.

D a grease molecule.

Question 7

What is a possible use of the compound with the structure shown?

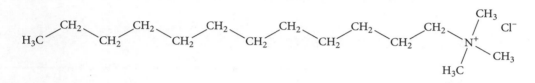

A To make biodiesel

B To make polymers

C To make tri-esters

D As fabric and hair conditioners

Question 8

What is the name given to the process shown below?

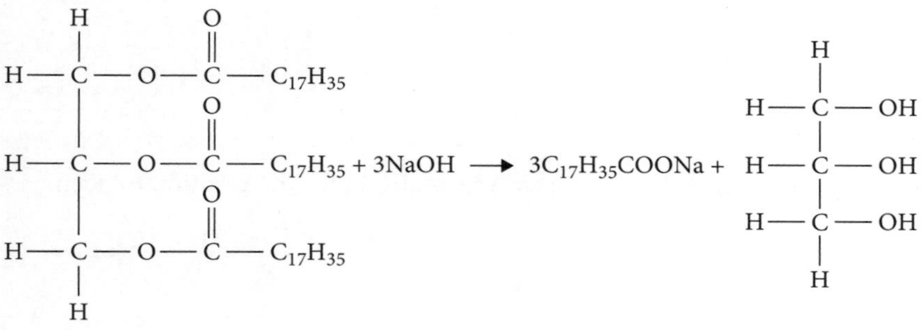

$$H-\overset{\overset{\displaystyle H}{|}}{\underset{\underset{\displaystyle H}{|}}{C}}-O-\overset{\overset{\displaystyle O}{||}}{C}-C_{17}H_{35}$$ ... $+ 3NaOH \longrightarrow 3C_{17}H_{35}COONa + \quad H-C-OH$ chain giving glycerol

A Esterification

B Hydrolysis

C Neutralisation

D Saponification

Question 9

Hard water can contain Ca^{2+} and/or Mg^{2+}. Which alternative is best when washing clothes in hard water?

A Use a cationic detergent because it does not interact with the ions that cause water hardness.

B Use soap because it does not interact with the ions that cause water hardness.

C Use soap because it will precipitate out the ions that cause water hardness.

D Use a cationic detergent because it will precipitate out the ions that cause water hardness.

Question 10 ○■■

Which statement is correct about the acids listed in the table?

Acid	pK_a
Methanoic acid	3.75
Ethanoic acid	4.75
Propanoic acid	4.87
Butanoic acid	4.82

A Methanoic acid is the weakest acid.

B Butanoic acid is the strongest acid.

C Ethanoic acid is a stronger acid than methanoic acid.

D Propanoic acid is a weaker acid than methanoic acid.

Section II: Short-answer questions

Instructions to students
- Answer all questions in the spaces provided.

Question 11 (10 marks)

57.80 mL of butan-1-ol was refluxed with excess ethanoic acid in the presence of concentrated sulfuric acid.

a ○■■ Write a balanced equation using full structural formulae to show the reaction when butan-1-ol is refluxed with ethanoic acid in the presence of concentrated sulfuric acid.   3 marks

_____

_____

_____

b ○■■ Explain why concentrated sulfuric acid was required for the reaction to occur.   1 mark

_____

_____

c ■■■ Calculate the volume of the organic product produced. The densities at 25°C of butan-1-ol and the organic product are 0.81 g mL^{-1} and 0.88 g mL^{-1} respectively, and the yield is 70%.   4 marks

_____

_____

_____

_____

_____

_____

d [●●] Describe how the organic product could have been separated from the mixture.   2 marks

_____

_____

_____

_____

Question 12 (3 marks) [●●]

Phenol is used as a disinfectant and an antiseptic, and in mouthwash and throat lozenges. It has the molecular formula C_6H_5OH and its structural formula is shown here. Phenol can be classified as an acidic substance.

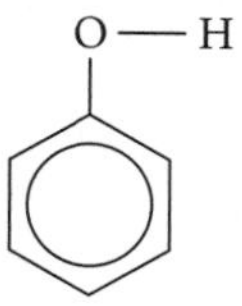

The pK_a values of phenol, ethanol and ethanoic acid are listed in the table.

Acid	pK_a
Ethanoic acid	4.75
Phenol	9.80
Ethanol	15.9

a Write an equation to demonstrate the acidity of phenol.   1 mark

b Compare the relative acid strengths of the substances listed in the table.   2 marks

_____

_____

_____

Question 13 (3 marks)

Ethanamine can be made in two steps:

I $CH_3CH_2Br + NH_3 \rightarrow CH_3CH_2NH_3^+Br^-$

II $CH_3CH_2NH_3^+Br^- + NH_3 \rightleftharpoons CH_3CH_2NH_2 + NH_4^+Br^-$

a [●●] Write a net equation for the production of ethanamine.   1 mark

_____

b [●●] Ethanamine is a gas at 25°C. Calculate the volume of ethanamine produced at 25°C if 12.6 L of ammonia is reacted.   2 marks

_____

_____

_____

_____

Question 14 (9 marks)

Ethanoic acid is produced when ethanamide undergoes hydrolysis with water and dilute hydrochloric acid. The corresponding ammonium salt is also produced.

a   ⬤◻◻   Write an equation to show the reaction.   1 mark

b   ⬤⬤◻   Ethanoic acid can also be produced from ethane. Suggest a pathway for this process.

Outline a series of steps, using structural formulae for the organic compounds and identifying the reagents and reaction conditions. There is no need to state reacting quantities.   4 marks

c   ⬤⬤⬤   In part **a**, 64 mL of 45% v/v ethanoic acid is produced.

Calculate the mass of ethanamide that was reacted, given the density of ethanoic acid is $1.05\,\mathrm{g\,mL^{-1}}$.   4 marks

Test 10: Polymers

Section I: 10 marks. Section II: 25 marks. Total marks: 35.
Suggested time: 60 minutes

Section I: Multiple-choice questions

Instructions to students
- For each question, circle the multiple-choice letter to indicate your answer.

Question 1

What is the name of the reaction that produces the substance shown? (The boxes represent carbon chains.)

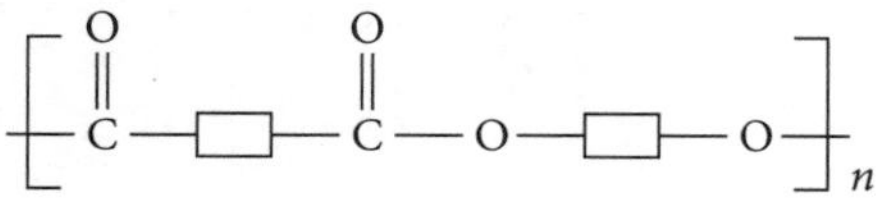

A Addition polymerisation

B Condensation polymerisation

C Esterification polymerisation

D Saponification

Question 2

A section of a polymer is shown.

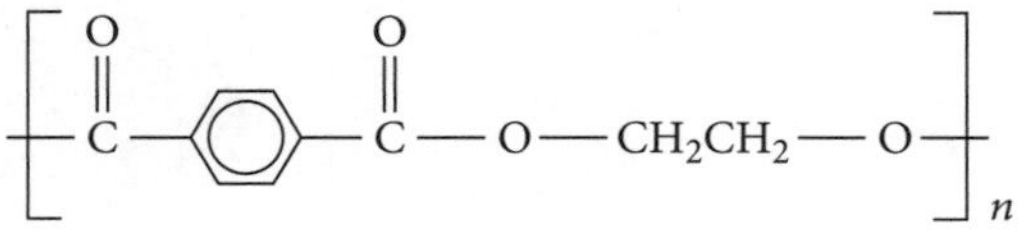

What is a possible name for the polymer?

A Polyester

B Polystyrene

C Low-density polyethylene (LDPE)

D Polyethylene terephthalate (PET)

Question 3

Which monomer shown can be used to make coatings for non-stick frying pans?

A

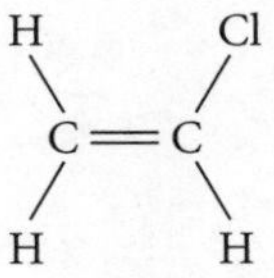

B

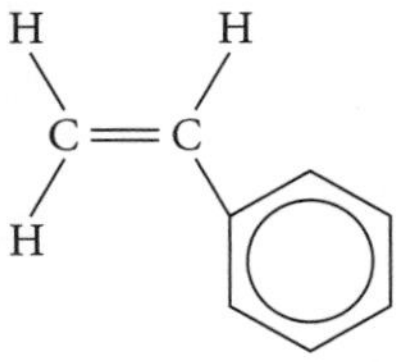

C

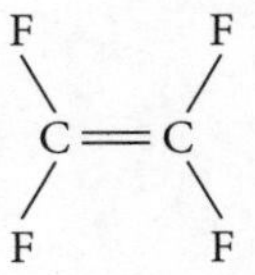

D

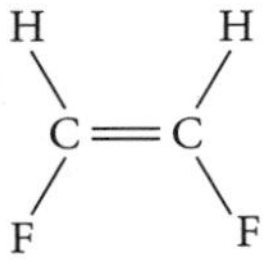

Question 4

What is the IUPAC name of the compound shown, which is used to make a polymer?

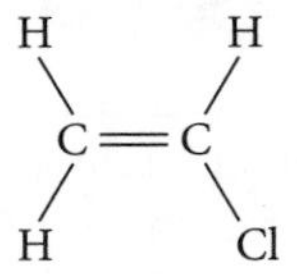

A Ethene

B Chloroethene

C Vinyl chloride

D Polyvinyl chloride

Question 5

Proteins are polymers that are made up of amino acids and are referred to as polypeptides. The structure of a polypeptide is shown.

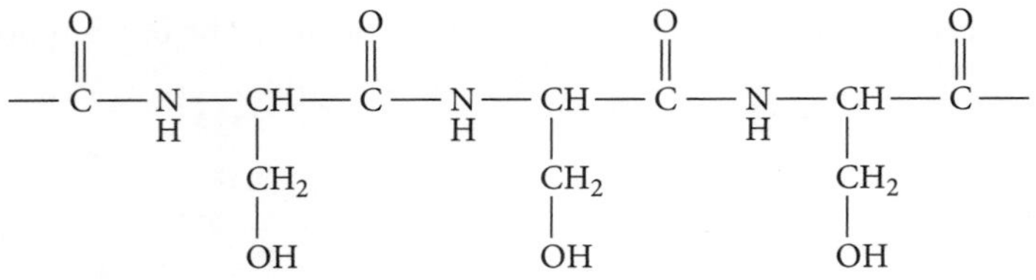

What is a possible structure for an amino acid in this part of the polypeptide?

A

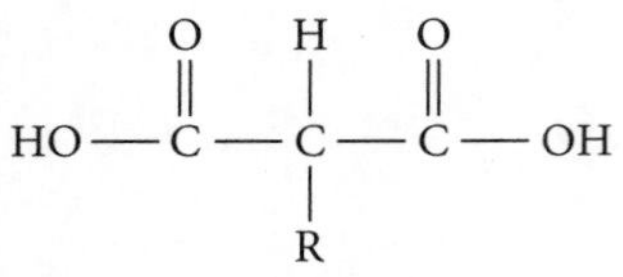

B

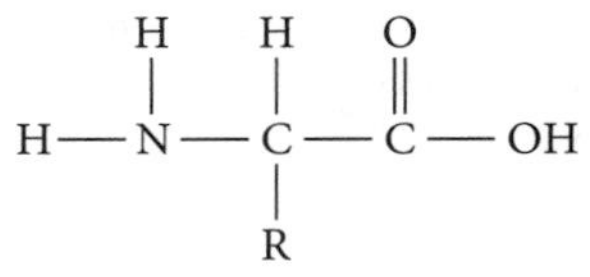

C

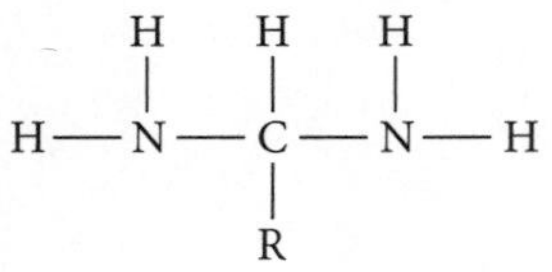

D

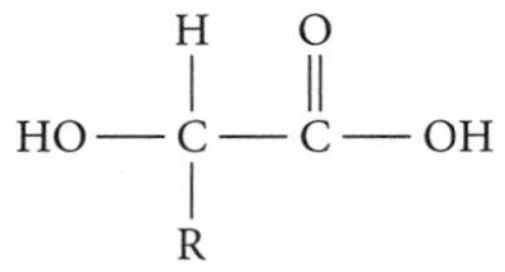

Use the following information to answer Questions 6 and 7.

The diagram shows a section of polyacrylonitrile, which can be used to make carpets.

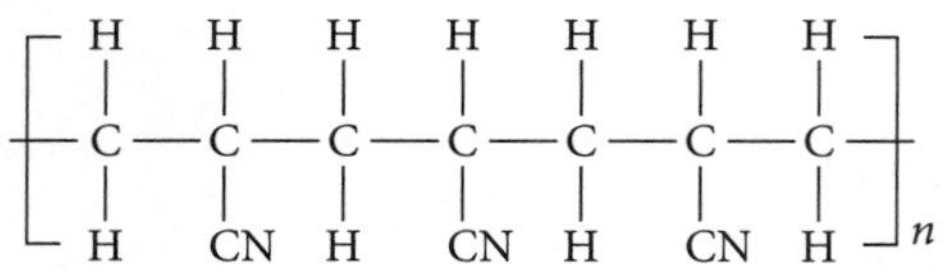

Question 6

What is the formula of the monomer used to make the polymer?

A

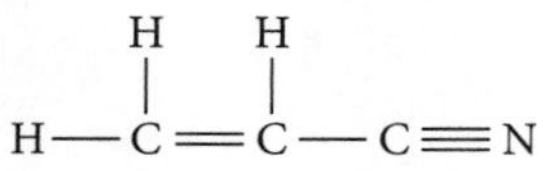

B

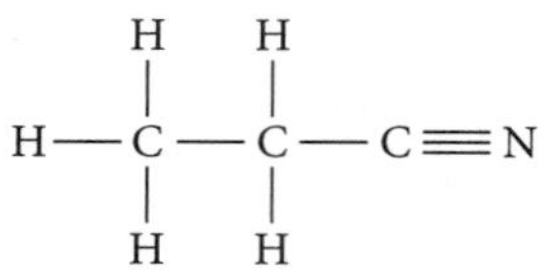

C

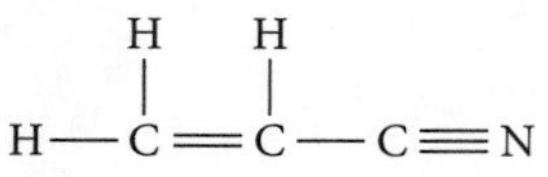

D

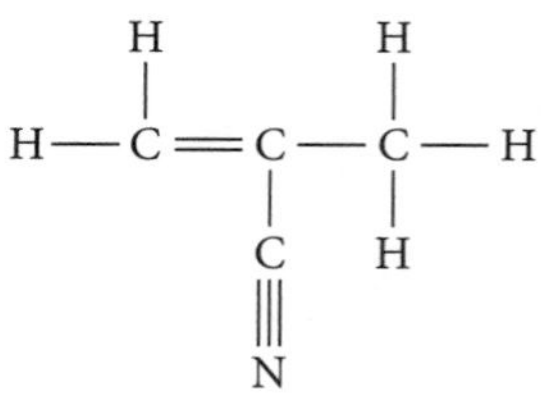

Question 7 ●●●

A sample of polyacrylonitrile has an average molar mass of $79\,596\,g\,mol^{-1}$. How many carbon atoms are in an average polymer molecule?

A 500

B 1500

C 3000

D 4500

Question 8 ●●

The tensile strength of a polymer can be defined as its resistance to breaking under tension.
Which alternative correctly identifies the tensile strengths of the listed polymers?

	Tensile strength (MPa)		
	70	**55**	**15**
A	Nylon	HDPE	PET
B	PET	HDPE	Nylon
C	Nylon	PET	HDPE
D	HDPE	Nylon	PET

Question 9 ●●●

The graph shows the distribution of molecular masses in a polymer sample.

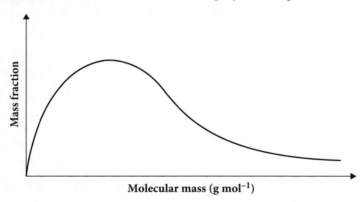

What is a possible identity of the polymer?

A Low-density polyethylene (LDPE)

B Polyester

C Nylon

D Not enough information is provided.

Question 10 〔●●〕

A section of a starch polymer is shown. It is formed by condensation polymerisation of glucose $(C_6H_{12}O_6)$ molecules.

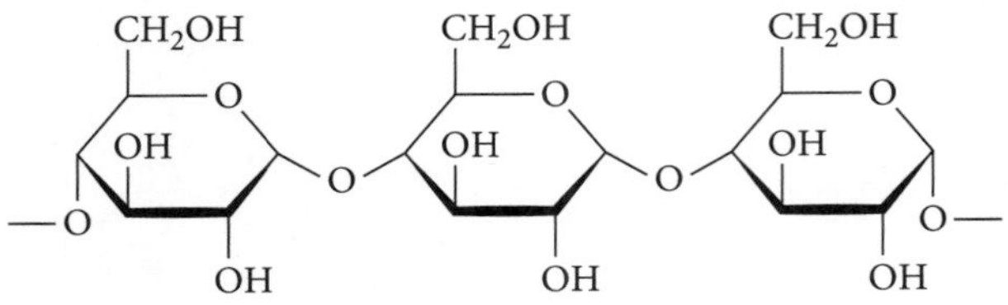

What is the approximate mass of a starch polymer made up of 10 000 glucose monomers?

A 1 621 418 g

B 1 711 480 g

C 1 783 562 g

D 1 801 560 g

Section II: Short-answer questions

> **Instructions to students**
> · Answer all questions in the spaces provided.

Question 11 (4 marks)

a 〔●●●〕 Match the substances polyethene, ethene, polyester and nylon 6,6 to the following melting points.   2 marks

−169°C _____

110°C _____

260°C _____

269°C _____

b 〔●●●〕 Explain your choices for the substances with the lowest and highest melting points. Include a diagram in your response.   2 marks

_____

_____

_____

_____

Question 12 (3 marks)

a ⬤⬤⬤ Draw structural formulae for the monomer(s) that form the polymer shown.   2 marks

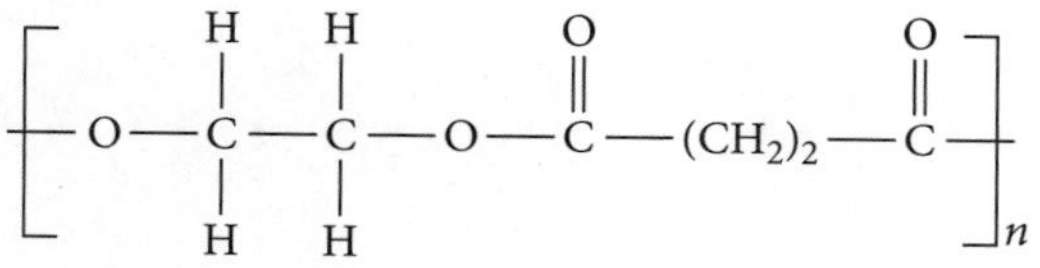

b ⬤◻◻ State the type of polymerisation reaction that forms the polymer in part **a**.   1 mark

_____

Question 13 (2 marks) ⬤◻◻

Explain why alkenes are desirable monomers for polymerisation reactions. In your response, refer to any other reagent that may be required in the process.

_____

_____

Question 14 (2 marks) ⬤◻◻

Compare two physical properties of LDPE and HDPE with reference to their structures.

_____

_____

Question 15 (4 marks) ⬤⬤⬤

The density of polytetrafluoroethylene (PTFE) is $2.2\,g\,mL^{-1}$. Calculate the number of fluorine atoms in a $1.5 \times 10^6\,L$ sample.

_____

_____

_____

_____

_____

Question 16 (5 marks) ⬤⬤⬜

Discuss the following statement in terms of the use of these polymers in society.

> Polymers have been useful in society. However, in recent times, the single-use plastic in shopping bags and straws has been replaced with renewable materials. Yet non-stick frying pans continue to be produced.

_____

_____

_____

_____

_____

_____

_____

_____

_____

Question 17 (5 marks)

Read the blog about the use of cable ties on the NASA $2.7 billion Perseverance rover.

Cable ties hold together the Perseverance rover in space.

The cable ties, which are made from ethylene tetrafluoroethylene, ETFE, also known as Tefzel, are used to secure the rover's interior and exterior conduits and components as well as research equipment.

The cable ties are resistant to chemicals, corrosion, radiation, extreme temperatures, weathering and ageing.

ETFE is odourless and non-toxic.

ETFE is made by reacting ethene (ethylene) with tetrafluoroethene, as shown in the equation below.

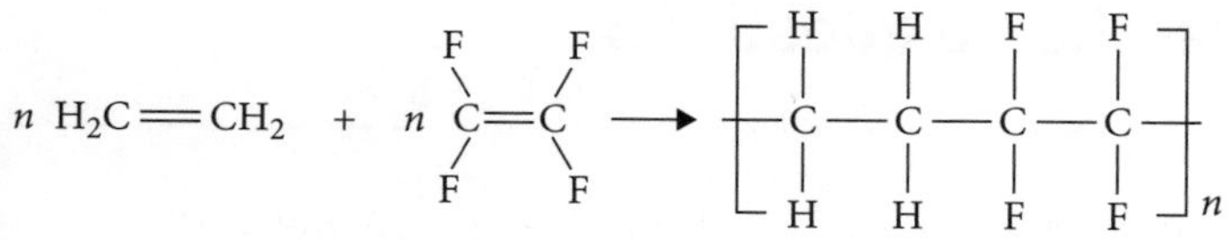

$$n\ H_2C{=}CH_2\ +\ n\ \underset{\underset{\displaystyle F}{|}}{\overset{\overset{\displaystyle F}{|}}{C}}{=}\underset{\underset{\displaystyle F}{|}}{\overset{\overset{\displaystyle F}{|}}{C} \longrightarrow \left[\begin{matrix} H & H & F & F \\ | & | & | & | \\ C{-}C{-}C{-}C \\ | & | & | & | \\ H & H & F & F \end{matrix} \right]_n$$

---

a Outline the significance of the use of cable ties on the Perseverance rover. **2 marks**

_____

_____

_____

_____

b At 25°C and 10 kPa, 10.00 ML of each of ethene and tetrafluoroethene were obtained for the above reaction.

Calculate the mass of ETFE that could be produced. Express your answer in tonnes, where 1 tonne = 10^6 g. **3 marks**

_____

_____

_____

_____

_____

_____

CHAPTER 4
MODULE 8: APPLYING CHEMICAL IDEAS

Test 11: Analysis of inorganic substances

Section I: 10 marks. Section II: 31 marks. Total marks: 41.
Suggested time: 74 minutes

Section I: Multiple-choice questions

Instructions to students
- For each question, circle the multiple-choice letter to indicate your answer.

Question 1

Which method listed is a safe test for determining whether lead(II) ions are present in a sample?

A Flame test

B Gravimetric analysis

C Addition of sodium nitrate solution

D Addition of sodium sulfate solution

Use the following information to answer Questions 2 and 3.

Two of the bonds in the following structure are labelled 1 and 2.

$$\left[\begin{array}{c} H \\ | \\ H-N-\overset{1}{\underset{|}{Ag}}-N-\overset{2}{H} \\ | \\ H \end{array} \right]^{+}$$

Question 2

What is the best description of the NH_3 groups in the ion?

A Complex ions

B Gases

C Ligands

D Molecules

Question 3

Which pair of terms for the bonds in the structure is correct?

	Bond 1	Bond 2
A	Ionic	Ionic
B	Ionic	Covalent
C	Covalent	Coordinate covalent
D	Coordinate covalent	Covalent

Question 4 ⬤⬤▢

What is the best way to distinguish between aqueous solutions of sodium carbonate and sodium chloride?

A AAS

B Colourimetry

C Flame test

D Addition of nitric acid

Question 5 ⬤⬤⬤

A student collected the following data when determining the percentage of sulfate in a commercial fertiliser. They crushed the fertiliser, and then added hydrochloric acid, followed by barium chloride solution. The student then filtered the barium sulfate precipitate and dried it in an oven overnight.

$$m(\text{fertiliser}) = 1.50\,g \qquad c(\text{HCl}) = 0.100\,\text{mol L}^{-1}$$

$$V(\text{HCl}) = 50.0\,mL \qquad c(\text{BaCl}_2) = 0.100\,\text{mol L}^{-1}$$

$$V(\text{BaCl}_2) = 100\,mL \qquad m(\text{BaSO}_4) = 4.12\,g$$

What is the best explanation for the student's results?

A The calculated mass of sulfate in the fertiliser is more than the mass of the fertiliser because the precipitate may not have been rinsed with water sufficiently and then not dried to constant mass.

B The calculated mass of sulfate in the fertiliser is more than the mass of the fertiliser because some of the precipitate may have been lost during filtering and then not dried to constant mass.

C The calculated mass of sulfate in the fertiliser is less than the mass of the fertiliser because the precipitate may not have been rinsed with water sufficiently and then not dried to constant mass.

D The calculated mass of sulfate in the fertiliser is less than the mass of the fertiliser because some of the precipitate may have been lost during filtering and then not dried to constant mass.

Question 6 ⬤⬤▢

The UV–visible absorbance spectra for pure water and samples labelled 1, 2 and 3 are given. The samples are ground water, tap water and river water. Ground water is the least contaminated and river water is the most contaminated.

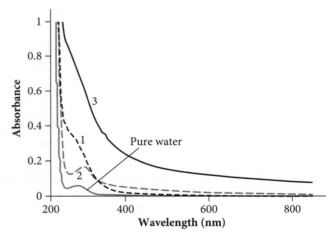

Which alternative correctly identifies the spectrum of each substance?

	Ground water	Tap water	River water
A	1	2	3
B	2	1	3
C	3	2	1
D	3	1	2

Question 7 ⬤⬤

The UV–visible spectrum of a solution is shown.

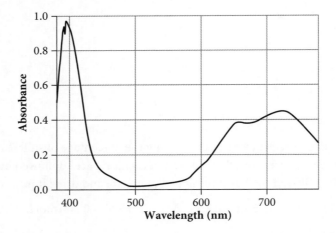

What colour is the solution?

A Blue

B Green

C Orange

D Red

Question 8 ⬤⬤

The following UV–visible spectrum is of a sample containing a mixture of solutions 1 and 2.

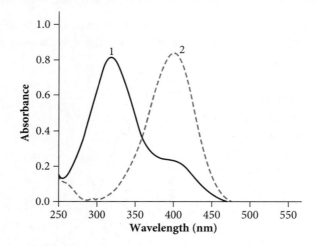

What wavelength could be used to measure the absorbance of solution 2 in the mixture?

A 300 nm

B 350 nm

C 400 nm

D 450 nm

Question 9 ●●●

AAS was used to analyse solutions containing lead(II) ions. A standard solution of 20 ppm lead(II) ions had an absorbance of 0.300. A solution of unknown concentration of lead(II) ions had an absorbance of 0.600.

When 200 mL of the unknown solution was reacted with excess sodium iodide solution, a precipitate formed. It was dried to constant mass and weighed. What is the mass of the precipitate?

A 17.8 g

B 17.8 mg

C 89.0 g

D 89.0 mg

Question 10 ©NESA 2019 SI Q20   ●●●

The manganese content in a 12.0-gram sample of steel was determined by measuring the absorbance of permanganate (MnO_4^-) using the following process.

The steel sample was dissolved in nitric acid and the $Mn^{2+}(aq)$ ions produced were oxidised to $MnO_4^-(aq)$ by periodate ions, $IO_4^-(aq)$, according to the following equation:

$$2Mn^{2+}(aq) + 5IO_4^-(aq) + 3H_2O(l) \rightarrow 2MnO_4^-(aq) + 5IO_3^-(aq) + 6H^+(aq)$$

The resulting solution was made up to a volume of 1.00 L, then 20.0 mL of this solution was diluted to 100.0 mL. The absorbance at 525 nm of the resulting solution was 0.50. A calibration curve for $MnO_4^-(aq)$ was constructed and is shown below.

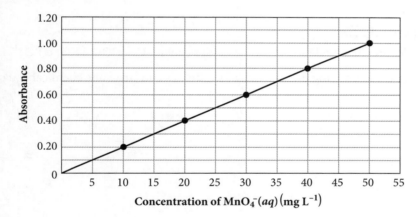

What was the percentage by mass of manganese in the steel sample?

A 0.019%

B 0.096%

C 0.48%

D 1.0%

Section II: Short-answer questions

Instructions to students
· Answer all questions in the spaces provided.

Question 11 (5 marks)

Analyse the need to monitor the environment with reference to specific ions and the analytical techniques that may be used.

_____

_____

_____

_____

_____

_____

_____

_____

Question 12 (5 marks)

Three reagent bottles are missing their labels. The bottles contain CH_3COO^-, I^- and PO_4^{3-} ions.

Outline a procedure for identifying the contents of the bottles and write one chemical equation for one test.

_____

_____

_____

_____

_____

_____

_____

Question 13 (3 marks) ●●

Silver nitrate is used to identify the presence of halide ions. The solution is first acidified with nitric acid to prevent the precipitation of non-halide ions such as carbonate. Dilute or concentrated ammonia solution is then added to confirm the presence of the halides.

Using complexation reactions, draw a flow chart to show how three solutions containing chloride, bromide and iodide may be distinguished.

Question 14 (5 marks) ●●●

To determine the sodium chloride concentration in tomato sauce, a precipitation titration was performed.

25.00 mL of the tomato sauce was mixed with distilled water to give a final volume of 250 mL. To 25.0 mL of this solution was added 50.0 mL of a 0.101 mol L^{-1} silver nitrate solution. Iron(III) ions were added to the solution and a titration was performed using 0.105 mol L^{-1} sodium thiocyanate. 38.72 mL of titrant was required to reach end point.

Calculate the concentration of sodium chloride in the tomato sauce in parts per million. Assume the density of the sauce was 1.20 g mL^{-1}.

Question 15 (7 marks)

The following table lists the absorbances of some standard solutions of mercury.

Sample	Concentration (ppm)	Absorbance
1	0.00	0.000
2	1.00	0.169
3	2.00	0.340
4	3.00	0.479
5	4.00	0.650
6	5.00	0.830

a 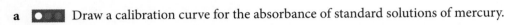 Draw a calibration curve for the absorbance of standard solutions of mercury.   3 marks

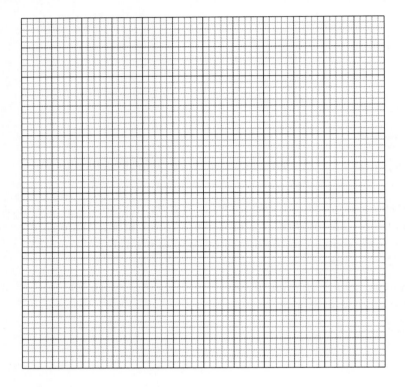

b Calculate the concentration of mercury in a sample with absorbance of 0.900.   1 mark

c Evaluate the reliability and validity of your response to part **b**.   3 marks

Test 12: Analysis of organic substances

Section I: 10 marks. Section II: 25 marks. Total marks: 35.
Suggested time: 60 minutes

Section I: Multiple-choice questions

Instructions to students
- For each question, circle the multiple-choice letter to indicate your answer.

Use the following spectrum to answer Questions 1 and 2.

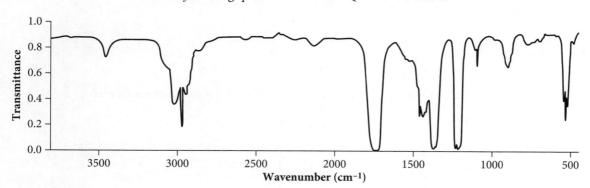

Question 1 ⬤◯◯

What is the name of the type of spectrum shown?

A Infrared spectrum

B Mass spectrum

C NMR spectrum

D UV–visible spectrum

Question 2 ⬤◯◯

What is the most likely structure of the compound shown in the spectrum?

A

$$H-\overset{\overset{\displaystyle H}{|}}{\underset{\underset{\displaystyle H}{|}}{C}}-\overset{\overset{\displaystyle O}{\parallel}}{C}\overset{}{\underset{\displaystyle O-H}{}}$$

B

$$H-\overset{\overset{\displaystyle H}{|}}{\underset{\underset{\displaystyle H}{|}}{C}}-\overset{\overset{\displaystyle \overset{\displaystyle H}{|}O}{|}}{\underset{\underset{\displaystyle H}{|}}{C}}-\overset{\overset{\displaystyle H}{|}}{\underset{\underset{\displaystyle H}{|}}{C}}-H$$

C

$$H-\overset{\overset{\displaystyle H}{|}}{\underset{\underset{\displaystyle H}{|}}{C}}-\overset{\overset{\displaystyle O}{\parallel}}{C}-\overset{\overset{\displaystyle H}{|}}{\underset{\underset{\displaystyle H}{|}}{C}}-H$$

D

$$H-\overset{\overset{\displaystyle H}{|}}{\underset{\underset{\displaystyle H}{|}}{C}}-\overset{\overset{\displaystyle H}{|}}{\underset{\underset{\displaystyle H}{|}}{C}}-\overset{\overset{\displaystyle H}{|}}{\underset{\underset{\displaystyle H}{|}}{C}}-O-H$$

Question 3 ●●○

Citric acid is found in lemon juice. Its structural formula is shown.

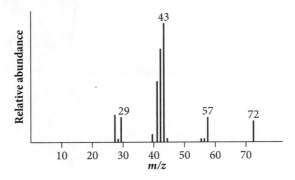

Which alternative correctly identifies whether citric acid reacts with the reagents listed?

	Acidified KMnO$_4$	HBr(g)	Na$_2$CO$_3$(aq)
A	Yes	Yes	Yes
B	No	No	Yes
C	No	Yes	Yes
D	Yes	No	No

Use the following mass spectrum of a hydrocarbon to answer Questions 4 and 5.

Question 4 ●●○

What is the likely formula of the compound?

A C$_5$H$_{12}$

B C$_6$H$_{12}$

C C$_3$H$_4$O$_2$

D C$_3$H$_5$NO

Question 5 ●○○

What is the *m/z* value of the base peak?

A 29

B 43

C 57

D 72

Question 6 ⬤⬤

The ^{13}C and ^1H NMR spectra of a compound are shown. The ^1H NMR spectrum shows the splitting pattern.

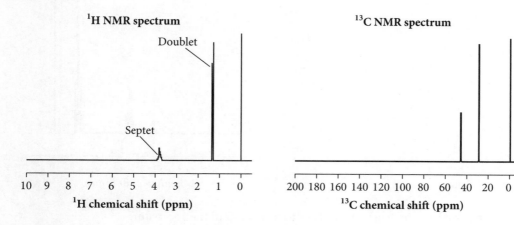

What is a possible identity of the organic compound?

A 1-Bromobutane

B 2-Bromobutane

C 1-Bromopropane

D 2-Bromopropane

Question 7 ⬤⬤⬤

The mass spectrum of an organic compound containing three carbon atoms, hydrogen atoms and chlorine atoms showed molecular or parent ion peaks at m/z 152, 150, 148 and 146. (Chlorine has isotopes with atomic masses 35 and 37.)

What is a possible identity of the organic compound?

A 1-Chloropropane

B 1,2-Dichloropropane

C 1,2,3-Trichloropropane

D Not enough information is provided.

Question 8 ●●■

The proton NMR spectrum of ethanol is shown.

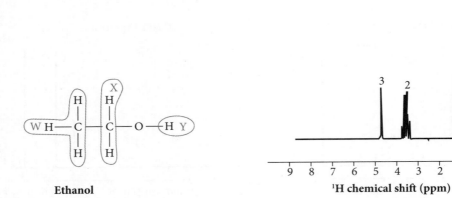

Ethanol

^{1}H chemical shift (ppm)

Which alternative correctly matches the protons to peaks 1–3 in the spectrum?

	Proton W	Proton X	Proton Y
A	1	2	3
B	3	2	1
C	2	3	1
D	1	3	2

Question 9 ●●●

Ibuprofen is a compound often found in medications used to treat pain and fever. Its structure is shown.

How many signals are in the ^{1}H NMR spectrum of ibuprofen, if the four hydrogen atoms on the benzene ring are ignored?

A 3

B 5

C 6

D 7

Question 10

13.304 mL of an unsaturated hydrocarbon with density $0.640\,\mathrm{g\,mL^{-1}}$ was reacted in the same mole ratio with 15.283 mL of bromine water that had a density of $1.307\,\mathrm{g\,mL^{-1}}$.

The ^{13}C NMR spectrum of the compound is shown.

What is the likely formula of the unsaturated hydrocarbon?

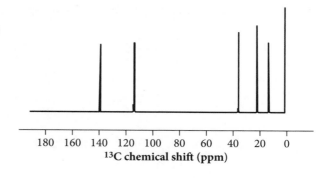

A C_5H_8

B C_5H_{10}

C C_6H_8

D C_6H_{10}

Section II: Short-answer questions

Instructions to students
- Answer all questions in the spaces provided.

Question 11 (8 marks)

Describe the processes used to analyse the structure of organic compounds. In your response, refer to specific functional groups or compounds.

Question 12 (6 marks)

a ◖●◗◗ Identify a solvent that is used in NMR spectroscopy and explain why it is useful.   3 marks

_____

_____

_____

_____

_____

_____

b ◖●●●◗ Explain what the horizontal scale shows in an NMR spectrum, with reference to why some peaks are further to the left than others.   3 marks

_____

_____

_____

_____

_____

Question 13 (4 marks) ◖●●◗

A student said infrared spectroscopy is the best spectroscopic procedure for distinguishing between ethanol, ethanoic acid and ethyl ethanoate. Evaluate the student's statement.

_____

_____

_____

_____

_____

_____

_____

Question 14 (7 marks) ●●●

Compound X was found to have 61% C, 15% H and 24% N. Compound X reacted with hydrochloric acid to produce an ammonium salt. To confirm the molecular structure of the compound, mass spectrometry, infrared spectroscopy, and ^{13}C and ^{1}H NMR spectrometry were performed. The resulting spectra are shown.

Mass spectrum

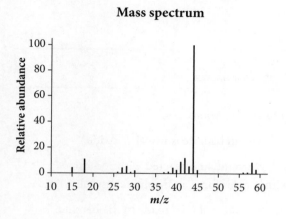

Infrared spectrum

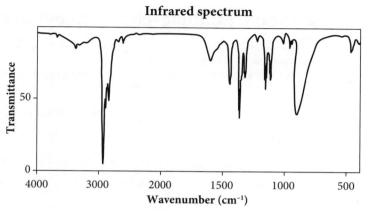

^{13}C NMR spectrum

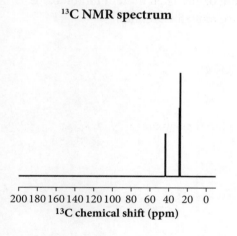

^{1}H NMR spectrum

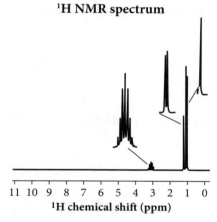

Signal position (ppm)	Integration	Splitting
3.1	1	7
1.2	6	2
1	2	1

Identify compound X. Justify your answer with references to the spectra.

_____

_____

_____

_____

_____

_____

_____

_____

_____

_____

_____

Test 13: Chemical synthesis and design

Section I: 10 marks. Section II: 22 marks. Total marks: 32.
Suggested time: 57 minutes

Section I: Multiple-choice questions

Instructions to students
- For each question, circle the multiple-choice letter to indicate your answer.

Use the following information to answer Questions 1–4.

Aluminium is extracted from the ore bauxite. The major impurity in bauxite is iron(III) oxide.

The Bayer process is used to extract aluminium. Bauxite is first ground and reacted with sodium hydroxide to produce soluble sodium aluminate, while the impurities form an insoluble residue.

An acid is not used in this process because it would dissolve the basic iron(III) oxide in the bauxite. This solution is then decomposed and reacted to produce aluminium oxide. Then the aluminium oxide is melted and electrolysed using the Hall–Heroult process to produce molten aluminium.

The overall reaction for the process is:

$$2Al_2O_3(s) + 3C(s) \rightarrow 4Al(s) + 3CO_2(g)$$

Question 1

Which statement describes a property of the aluminium oxide found in the bauxite?

A It is basic.

B It is amphoteric.

C It is amphiprotic.

D It is neutral.

Question 2

Which alternative describes what happens at the anode and cathode during the electrolysis process?

	Anode	Cathode
A	Aluminium ions are oxidised and gain electrons.	Carbon is reduced and loses electrons.
B	Aluminium ions are oxidised and lose electrons.	Carbon is reduced and gains electrons.
C	Carbon is oxidised and loses electrons.	Aluminium ions are reduced and gain electrons.
D	Carbon is oxidised and gains electrons.	Aluminium ions are reduced and lose electrons.

Question 3

What is the percentage purity of the bauxite if 2.25 tonnes of bauxite produces 750 kg of pure aluminium?
(1 tonne = 10^6 g)

A 15.8%

B 31.0%

C 47.2%

D 63.0%

Question 4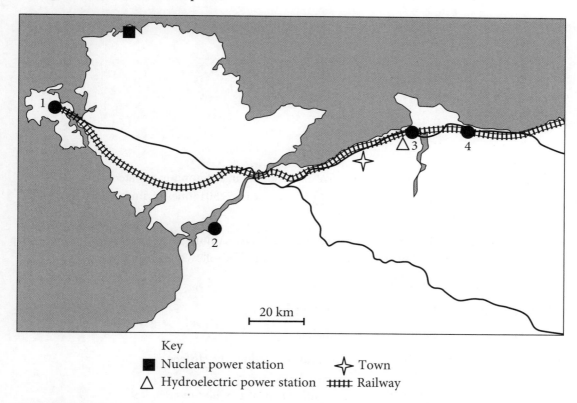

The map shows four possible locations for an aluminium extraction plant. Which statement is correct about the preferred location of the plant?

20 km

Key
■ Nuclear power station   ✦ Town
△ Hydroelectric power station   ⊞⊞⊞ Railway

A Location 1 is the preferred location because it is far away from a populated area but has road and railway access, and access to nuclear power for the electrolysis process.

B Location 2 is the preferred location because it is isolated and electricity can be transported by powerlines.

C Location 3 is the preferred location because it is close to the town where employees may live, and has road and railway access, and has access to hydroelectric power for the electrolysis process.

D Location 4 is the preferred location because it is somewhat isolated but has access to road, railway and hydroelectric power.

Question 5

Lacosamide is a medication used to treat epilepsy. It was first synthesised in 1996, in a three-step process as summarised below. The percentage by mass yield of each step is given:

$$\text{Cheap reactant} \xrightarrow{99\%} X \xrightarrow{37\%} Y \xrightarrow{80\%} \text{Lacosamide}$$

What mass of lacosamide would have been produced from 1.00 kg of the cheap starting material?

A 293 g

B 366 g

C 800 g

D Not enough information is provided.

Question 6 ⬤⬤

In a laboratory, 515.0 g of methanol was refluxed with 515.0 g of ethanoic acid and 15.0 g of concentrated sulfuric acid. The product was extracted and had a mass of 515.0 g.

What is the approximate percentage yield in this process?

A 43%

B 50%

C 81%

D 100%

Question 7 ⬤⬤

In August 2020, in Lebanon, there was a massive explosion in an ammonium nitrate storage facility in Beirut's shipping port district close to the city. It is claimed that 2750 tonnes of ammonium nitrate had been stored at this location for about six years.

Ammonium nitrate is used in fertilisers and to make explosives. Although the storage facility was next to the sea, the risk of an explosion could have been minimised by following which procedure listed?

A Regular monitoring of the ammonium nitrate

B Reducing the amount of ammonium nitrate stored

C Regular disposal of ammonium nitrate in the sea

D Storing the ammonium nitrate in large pellet form to reduce the surface area

Use the following information to answer Questions 8 and 9.

Green chemistry aims to develop chemical processes that reduce the use and production of hazardous substances. It aims to develop a sustainable approach of promoting processes that are environmentally friendly.

The term 'atom economy' is often used to refer to processes that minimise wastage. A high atom economy demonstrates a sustainable process. Atom economy is calculated as shown:

$$\% \text{ atom economy} = \frac{\text{relative molar mass of desired product} \times \text{ stoichiometric number}}{\text{sum of relative molar masses of products} \times \text{ appropriate stoichiometric number(s)}} \times 100$$

Question 8 ⬤⬤

Dichloromethane (CH_2Cl_2) and hydrogen chloride gas are formed when methane gas reacts with chlorine gas. What is the percentage atom economy of CH_2Cl_2?

A 11.16%

B 47.21%

C 53.80%

D 70.00%

Question 9 ⬤⬤⬤

Which of the following actions would improve the atom economy of dichloromethane production in the process described?

A Release the $HCl(g)$ into the atmosphere.

B Neutralise the $HCl(g)$ by adding $NaOH(aq)$.

C Bubble the $HCl(g)$ through water to produce useful hydrochloric acid.

D Monitor the production of $HCl(g)$, using moistened litmus, so that it is not released into the atmosphere.

Question 10

Which factor is **not** usually considered when planning the location of a chemical factory?

A Availability of energy

B Availability of sunlight

C Environment

D Transport

Section II: Short-answer questions

Instructions to students
- Answer all questions in the spaces provided.

Question 11 (6 marks) ©NESA 2011 SII Q30

The flow chart outlines the sequence of steps in the Ostwald process for the manufacture of nitric acid.

$$\text{Step 1} \quad 4NH_3(g) + 5O_2(g) \xrightleftharpoons{900°C} 4NO(g) + 6H_2O(g) \quad \Delta H = -950 \text{ kJ}$$

$$\downarrow$$

$$\text{Step 2} \quad 2NO(g) + O_2(g) \rightleftharpoons 2NO_2(g) \quad \Delta H = -114 \text{ kJ}$$

$$\downarrow$$

$$\text{Step 3} \quad 3NO_2(g) + H_2O(\ell) \rightarrow 2HNO_3(aq) + NO(g) \quad \Delta H = -117 \text{ kJ}$$

Explain the reaction conditions required at each step of the Ostwald process to maximise the yield and production rate of nitric acid.

_____

_____

_____

_____

_____

_____

_____

_____

_____

_____

_____

_____

Question 12 (6 marks) ●●

In Australia, $SO_2(g)$ produced by the extraction of metals is often used in the production of sulfuric acid in the contact process. For example, pentlandite ($Fe_9Ni_9S_8$) is a nickel sulfide ore, which is combusted in air to form sulfur dioxide:

$$Fe_9Ni_9S_8(s) + 17O_2(g) \rightarrow 8SO_2(g) + 9NiO(s) + 9FeO(s)$$

Assuming the yield is 70.0% at 290.0°C and 160.0 kPa, calculate the volume of sulfur dioxide produced when 1.88 tonnes of pentlandite is combusted in air under these conditions.
(1 tonne $= 10^6 g$)

Question 13 (10 marks)

The graph shows the percentage yield of ammonia under different reaction conditions.

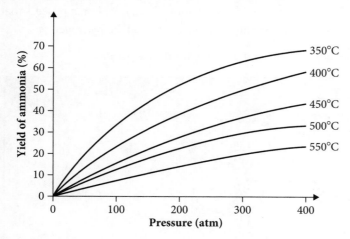

a ◖●▮▮ Identify the optimum conditions for the maximum yield of ammonia.   2 marks

_____

_____

b ◖●●▮ Calculate the masses of ammonia produced at 450°C and 200 atm when 1.50 kg of nitrogen gas is reacted with 700.0 g of hydrogen gas according to the reaction:

$$N_2(g) + 3H_2(g) \rightleftharpoons 2NH_3(g)$$ 4 marks

_____

_____

_____

_____

_____

_____

_____

_____

_____

c   The traditional source of reactants for the Haber process was nitrogen from the atmosphere and hydrogen from the reaction of steam with hydrocarbons.

However, in recent years the production of 'green ammonia' has relied on a process that minimises CO_2 production and uses electrolysis of water to produce hydrogen gas.

The map below shows four possible sites for a 'green ammonia' plant.

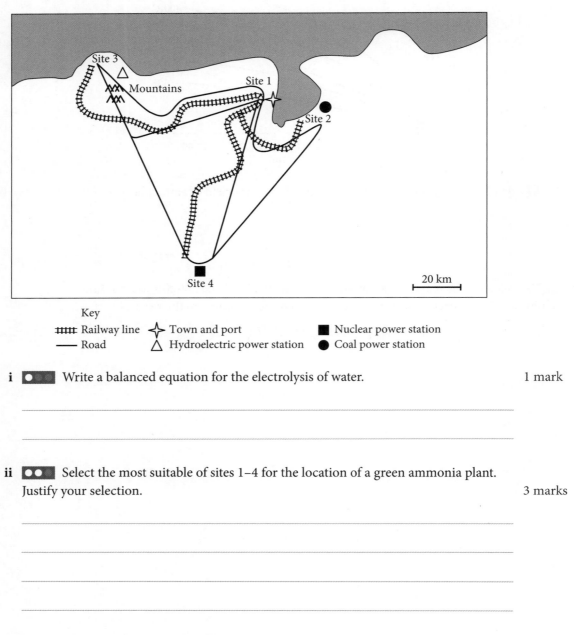

Key

ⅠⅠⅠⅠ Railway line   ✦ Town and port   ■ Nuclear power station
— Road   △ Hydroelectric power station   ● Coal power station

i   ◐◑◑ Write a balanced equation for the electrolysis of water.   1 mark

ii ◐◐◑ Select the most suitable of sites 1–4 for the location of a green ammonia plant. Justify your selection.   3 marks

Chemistry

PRACTICE HSC EXAM 1

General instructions
- Reading time – 5 minutes
- Working time – 3 hours
- Write using black pen
- Draw diagrams using pencil
- Calculators approved by NESA may be used
- A formulae sheet, data sheet and Periodic Table are provided at the back of this paper

---

Total marks: 100

Section I – 20 marks
- Attempt Questions 1–20
- Allow about 35 minutes for this section

Section II – 80 marks
- Attempt Questions 21–35
- Allow about 2 hours and 25 minutes for this section

Section I

20 marks
Attempt Questions 1–20
Allow about 35 minutes for this section
Circle the correct multiple-choice option for Questions 1–20.

Question 1

What is the IUPAC name of the structure shown?

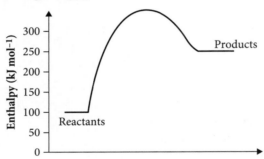

A *N,N*-Dimethylmethanamine

B *N,N,N*-Trimethylmethanamine

C *N,N,N*-Trimethylpropanamine

D Tetramethanamine

Question 2

A chemical reaction displays the following energy profile.

What is the enthalpy of the forward reaction?

A $+250\,\text{kJ}\,\text{mol}^{-1}$

B $+150\,\text{kJ}\,\text{mol}^{-1}$

C $-150\,\text{kJ}\,\text{mol}^{-1}$

D $-100\,\text{kJ}\,\text{mol}^{-1}$

Question 3

The equipment shown was used to separate the product of a reaction in organic chemistry.

What were the probable reactants?

A Ethanol + acidified potassium dichromate

B Ethanol + sodium metal

C Ethanol + bromine

D Ethanol + ethanoic acid + concentrated sulfuric acid

Question 4

What are the correct names for the acids HBr, HBrO, $HBrO_2$ and $HBrO_3$?

	HBr	HBrO	$HBrO_2$	$HBrO_3$
A	Bromic acid	Bromous acid	Hypobromous acid	Hydrobromic acid
B	Hydrobromic acid	Hypobromous acid	Bromous acid	Bromic acid
C	Bromous acid	Bromic acid	Hydrobromic acid	Hypobromous acid
D	Hypobromous acid	Bromic acid	Hydrobromic acid	Bromous acid

Question 5

How many isomers have the formula $C_4H_{10}O$?

A 2

B 3

C 4

D 5

Question 6

At 10°C, the ionisation constant of water is 2.93×10^{-15}. What is water's pOH?

A 6.73

B 7

C 7.27

D 14

Question 7

Silver nitrate solution is used to test for the presence of halides Cl^-, Br^- and I^-. However, although the silver precipitates of the halides are slightly different colours, it is often necessary for a further test to be carried out.

Which alternative is correct for the confirmatory test after a precipitate forms with acidified silver nitrate solution?

	Test	Type of reaction
A	Add dilute ammonia solution. If the precipitate dissolves, it is Br^-. If the precipitate does not dissolve, add concentrated ammonia. If the precipitate now dissolves, it is I^-.	Precipitation
B	Add dilute ammonia solution. If the precipitate dissolves, it is Br^-. If the precipitate does not dissolve, add concentrated ammonia. If the precipitate now dissolves, it is I^-.	Complexation
C	Add dilute ammonia solution. If the precipitate dissolves, it is Cl^-. If the precipitate does not dissolve, add concentrated ammonia. If the precipitate now dissolves, it is Br^-.	Precipitation
D	Add dilute ammonia solution. If the precipitate dissolves, it is Cl^-. If the precipitate does not dissolve, add concentrated ammonia. If the precipitate now dissolves, it is Br^-.	Complexation

Question 8

An aqueous sample contained either sulfate or phosphate ions. A student was trying to determine if it contained phosphate ions by adding barium chloride ions. A precipitate was observed and the student concluded that the sample contained phosphate ions.

Which statement is true about the student's conclusion?

A   It is valid.

B   It is not valid because the sample should have been made acidic first to ensure only barium phosphate precipitated.

C   It is not valid because the sample should have been made basic first to ensure only barium phosphate precipitated.

D   It is not possible to use barium chloride to distinguish between sulfate and phosphate ions.

Question 9

Which row of the table correctly identifies what happens when the compound shown is reacted with acidified potassium dichromate solution?

	Observation	Organic product formed
A	No visible reaction	—
B	Orange colour turns green	2,2-Dimethylpropanoic acid
C	Orange colour turns green	2-Methyl-2-ethylethanoic acid
D	Purple colour turns colourless	2-Dimethylpropanoic acid

Question 10 ●●

How many peaks would be observed in a ^{13}C NMR analysis of ethyl acetate, also referred to as ethyl ethanoate?

A   4

B   3

C   2

D   1

Question 11 ●●

In which version of the following reaction are the conjugate acid–base pairs correctly identified?

A $NH_3(aq) + HCl(g) \rightarrow NH_4Cl(s)$
    base   acid   base

B $NH_3(aq) + HCl(g) \rightarrow NH_4Cl(s)$
    base   acid   acid

C $NH_3(aq) + HCl(g) \rightarrow NH_4Cl(s) + H_2O(l)$
    acid   acid   base   acid

D $NH_3(aq) + HCl(g) \rightarrow NH_4Cl(s) + H_2O(l)$
    acid   base   acid   base

Question 12 ●●

Which row of the table correctly identifies the changes in entropy and enthalpy that occur during photosynthesis?

	Entropy	Enthalpy
A	Increase	Exothermic
B	Decrease	Exothermic
C	Increase	Endothermic
D	Decrease	Endothermic

Question 13 ●●

What is the correct equilibrium constant expression for iron(III) thiocyanate at equilibrium?

A $\dfrac{[Fe^{3+}][SCN^-]}{[Fe(SCN)^{2+}]}$

B $\dfrac{[Fe(SCN)^{2+}]}{[Fe^{3+}][SCN^-]}$

C $\dfrac{Fe(SCN)^{2+}(aq)}{Fe^{3+} + SCN^-(aq)}$

D $\dfrac{Fe^{3+} + SCN^-(aq)}{Fe(SCN)^{2+}(aq)}$

Question 14 ●●

What is the function of the magnet in a mass spectrometer?

A   To cause the ionisation of positively charged particles

B   To attract negative and positive ions to their corresponding poles for detection

C   To sort the ions by mass using acceleration and deflection

D   To remove neutral particles from the sample

Question 15 ⬤⬤

Which graph represents the conductometric titration of hydrobromic acid and calcium hydroxide?

A
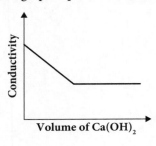

B
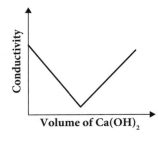

C
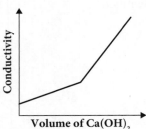

D
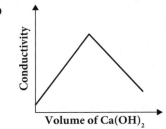

Question 16 ⬤⬤⬤

A student makes a solution with a final volume of 200 mL by mixing 100 mL of 0.0500 mol L^{-1} barium nitrate solution with 100 mL of 0.100 mol L^{-1} sodium hydroxide solution.

Which row of the table correctly identifies whether a precipitate will form under these conditions, and the reason?

	Will a precipitate form?	Reason
A	Yes	$Q > K_{sp}$
B	Yes	$Q < K_{sp}$
C	No	$Q > K_{sp}$
D	No	$Q < K_{sp}$

Question 17 ⬤⬤⬤

Which of the following is the positional isomer of methyl propanoate?

$$H_3C - CH_2 - \overset{\overset{\textstyle O}{\|}}{C} - O - CH_3$$

Methyl propanoate

A   Ethyl acetate

B   Ethyl propanoate

C   Butanoic acid

D   Butan-2-ol

Question 18 ⬤⬤⬤

What is the final pH when 100 mL of 0.10 M CH$_3$COOH with a K_a of 1.5×10^{-5} is added to 50 mL of 0.10 M of NaOH(aq)?

A   1.47

B   1.477

C   3.15

D   3.152

Question 19 ⬤⬤⬤

The following information is given about a compound. Compound W:

- has two signals in its ^{13}C NMR spectrum

- has two singlet signals in its ^{1}H NMR spectrum in the ratio 9:1

- reacts with HBr(g) to produce compound X

- reacts with concentrated sulfuric acid to produce compound Y, which subsequently reacts with HBr(g) to produce compound X

- has no visible reaction with acidified potassium dichromate or potassium permanganate.

Which alternative correctly identifies compounds W, X and Y?

	Compound W	Compound X	Compound Y
A	Propan-1-ol	1-Bromopropane	Propene
B	Propan-2-ol	2-Bromopropane	Prop-2-ene
C	2-Methylpropan-1-ol	1-Bromo-2-methylpropane	2-Methylpropene
D	2-Methylpropan-2-ol	2-Bromo-2-methylpropane	2-Methylpropene

Question 20 ⬤⬤⬤

Studies show that the use of copper in gym equipment significantly reduces bacterial growth on the surfaces of the equipment. A 5.0 g sample of a dumbbell was dissolved in concentrated nitric acid to produce copper ions. The resulting solution was made up to 1 L. Then 10.0 mL of this solution was diluted to 500 mL. The absorbance reading of this diluted sample was 0.015.

A calibration curve for Cu^{2+} was constructed and is shown below.

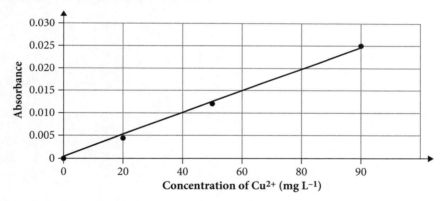

What is the percentage by mass in the dumbbell sample?

A 0.6%

B 1.2%

C 12%

D 60%

Section II

80 marks
Attempt Questions 21–35
Allow about 2 hours and 25 minutes for this section

Instructions
- Answer the questions in the spaces provided. These spaces provide guidance for the expected length of response.
- Show all relevant working in questions involving calculations.
- Extra writing space is provided at the back of this booklet. If you use this space, clearly indicate which question you are answering.

Question 21 (4 marks)

The table shows the pH of water at different temperatures.

Temp. (°C)	pH
0	7.47
25	7.00
50	6.63
100	6.14

a   ☐☐☐ Write the equation for the ionisation of water.   1 mark

_____

b   ☐☐☐ Identify whether the process of ionisation of water is exothermic or endothermic. Explain your response with reference to the data provided.   3 marks

_____

_____

_____

_____

_____

End of Question 21

Question 22 (4 marks)

The following flow chart is for a reaction pathway beginning with hex-3-ene.

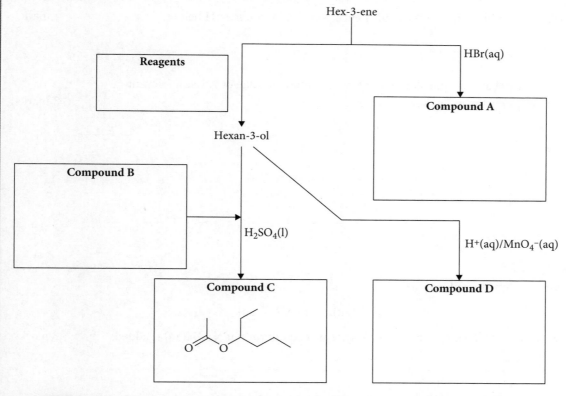

Hex-3-ene

Reagents

HBr(aq)

Compound A

Hexan-3-ol

Compound B

$H_2SO_4(l)$

$H^+(aq)/MnO_4^-(aq)$

Compound C

Compound D

a   In the relevant box on the flow chart, draw the structural formula of compound A.   1 mark

b   Identify the **two** reagents needed to convert hex-3-ene to hexan-3-ol. Write their chemical formulas in the relevant box on the flow chart.   1 mark

c   Compound B reacted with hexan-3-ol to produce compound C. Draw the structural formula of compound B in the relevant box on the flow chart.   1 mark

d   Draw the structural formula of compound D in the relevant box on the flow chart.   1 mark

Question 23 (5 marks)

Answer the following questions based on the titration curve below.

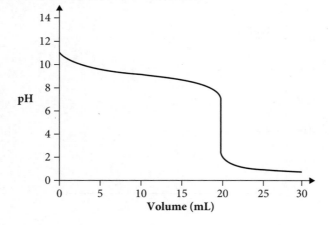

Question 23 continues on page 96

a   ⬤◯◻ Circle the correct options in the following sentence.   1 mark

This is a titration of a (strong base/weak base) with a (strong acid/weak acid).

b   ⬤◯◻ Draw a cross (X) on the curve at the equivalence point and write the pH below.   1 mark

_____

c   ◻◻◻ Explain why the pH at the equivalence point does not have a value of 7. Use a relevant
chemical equation to support your answer.   3 marks

_____

_____

_____

_____

Question 24 (6 marks) ⬤⬤◻

Consider the following reaction at equilibrium.

$$Fe^{3+}(aq) + SCN^-(aq) \rightleftharpoons FeSCN^{2+}(aq)$$

a   Describe and explain what will happen to the above system at equilibrium if NaOH(aq) is added.   3 marks

_____

_____

_____

_____

_____

_____

b   On the following graph, show the effect of adding NaOH(aq).   3 marks

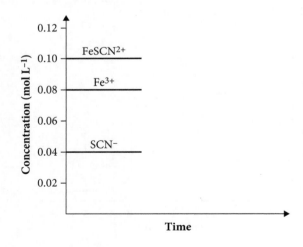

End of Question 24

Question 25 (3 marks) ⬤⬤

Infrared spectra 1–3 are of an alcohol, an alkane and an alkanoic acid.

Correctly identify which spectrum relates to which organic compound. Support your choice with reference to the spectra.

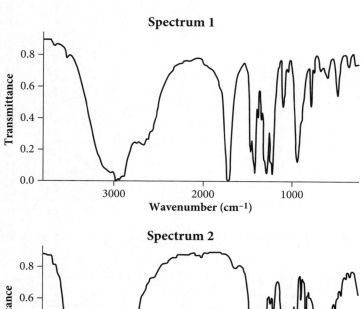

Spectrum 1

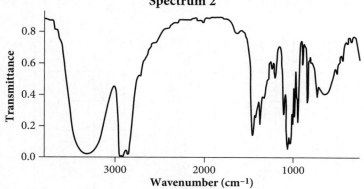

Spectrum 2

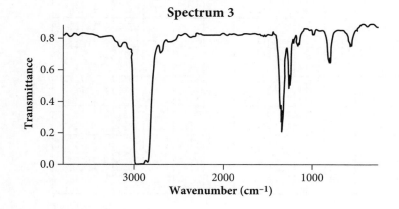

Spectrum 3

_____

_____

_____

_____

_____

_____

End of Question 25

Question 26 (7 marks) ●●

A 25.00 mL aliquot of phosphoric acid (H_3PO_4) was titrated
with 0.201 M sodium hydroxide (NaOH) solution. The pH values
for various volumes of sodium hydroxide are shown in the table.

V(NaOH) (mL)	pH
0	2.9
5.0	4.2
10.0	4.6
15.0	4.9
20.0	5.4
22.0	5.6
24.0	6.1
25.0	8.8
26.0	11.4
28.0	11.9
30.0	12.1
35.0	12.4
40.0	12.6
45.0	12.6

By graphing the data in the table and performing relevant calculations, determine the concentration of the
phosphoric acid solution.

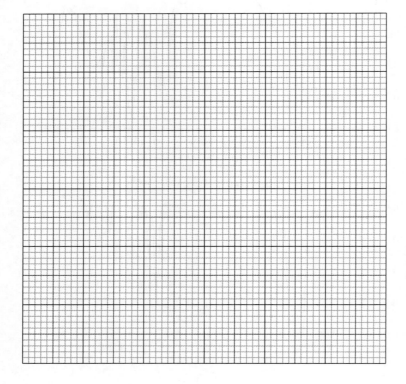

End of Question 26

Question 27 (5 marks) ⬤⬤⬤

Labels have fallen off six reagent bottles. The labels are:

- aqueous sodium acetate

- aqueous lead(II) nitrate

- ethanamide

- ethanamine

- ethanoic acid

- ethanol.

Outline a suitable procedure, using minimal chemicals, to identify the contents of each bottle.

End of Question 27

Question 28 (5 marks) ⬤⬤⬤

A water sample was analysed to determine the chloride ion content. 50.00 mL of this water was added to 100.00 mL of 0.1042 mol L^{-1} AgNO$_3$(aq). The excess silver nitrate was titrated against 0.05351 mol L^{-1} KSCN(aq). The titration was repeated several times and the average titre was 41.25 mL. Calculate the concentration of chloride ions in the water, expressed in mg L^{-1}, and state the species that was added to the titration to act as an indicator.

_____

_____

_____

_____

_____

_____

Question 29 (3 marks) ⬤⬤

Calculate the mass of ethanol that must be burned to increase the temperature of 155 g of water by 72°C, if exactly 80% of the heat released by this combustion is released to the surroundings. The theoretical molar heat of combustion of ethanol is 1367 kJ mol^{-1}.

_____

_____

_____

_____

_____

Question 30 (9 marks)

A student wanted to determine the concentration of a 250 mL solution of calcium hydroxide. After confirming the concentration of the secondary standard, phosphoric acid, to be 0.77 mol L^{-1}, the student used the following method to determine the concentration of calcium hydroxide.

1   Clean and rinse the burette with deionised water.

2   Condition the burette with a small volume of calcium hydroxide. Repeat this process three times.

3   Using a funnel, fill the burette with calcium hydroxide and record the initial volume.

4   Label five 100 mL conical flasks A–E.

5   Pour a 25.0 mL aliquot of phosphoric acid into the conical flask labelled A. Add three drops of bromocresol green.

6   Titrate the calcium hydroxide against the phosphoric acid until a colour change from green to aqua is seen. Record the volume of calcium hydroxide used.

7   Repeat steps 5 and 6 for flasks B–E.

Question 30 continues on page 101

The colour range of bromocresol green is shown below.

pH	Colour
3.5	Yellow
4.0	Light green
4.5	Green
5.0	Aqua
5.5	Light blue
6.0	Dark blue

a   Assess the suitability of using bromocresol green in this titration.   4 marks

The student recorded the following results.

Titre	Volume of calcium hydroxide (mL)
A	13.2
B	11.5
C	11.3
D	11.4
E	11.6
Average	11.4

b   Assess the reliability of the results obtained.   2 marks

c   Calculate the concentration of the calcium hydroxide.   3 marks

End of Question 30

Question 31 (9 marks)

The Haber process converts atmospheric nitrogen and hydrogen, in the presence of a solid iron catalyst, to ammonia. The catalyst is most effective at 400°C.

$$N_2(g) + 3H_2(g) \rightleftharpoons 2NH_3(g) \qquad \Delta H = -91.8 \, \text{kJ mol}^{-1}$$

A mixture of 4.00 mol of $N_2(g)$ and 4.00 mol of $H_2(g)$ was placed in a 3.00 L sealed reaction vessel and kept at 400°C until equilibrium was reached. At equilibrium, the vessel was found to contain 0.66 mol of $NH_3(g)$.

a   Calculate the equilibrium constant for the system at 400°C.   3 marks

b   The manufacturer wanted to increase the yield of ammonia produced. Evaluate the factors that need to be considered, such as reaction conditions, safety and cost, in order to viably increase yield.   6 marks

End of Question 31

Question 32 (4 marks) ⬤⬤⬤

An insoluble white ionic compound was provided to students for them to determine its identity.
The students performed several tests on the compound. Their results are shown in the table.

Compound added to a solution of	Observation
Barium nitrate	No visible change
Sodium hydroxide	No visible change
Silver nitrate	No visible change
Nitric acid	Bubbling

Because the bubbles indicated a chemical reaction with nitric acid, the students performed further tests on the remaining liquid.

Liquid added to a solution of	Observation
Sodium chloride	No visible change
Sodium hydroxide	No visible change
Sodium sulfate	White precipitate formed

Using these results, determine the identity of the insoluble white ionic compound. Justify your choice with appropriate net ionic equations.

End of Question 32

Question 33 (7 marks)

Fermentation is a process used to produce alcohols.

a   ⬛⬜⬜ Identify the fermentation conditions required in the production of an alcohol.   1 mark

b   ⬛⬛⬜ Outline how a named alcohol could be produced in a school laboratory.   3 marks

Microbe contamination in the fermentation vessel often leads to the formation of unwanted methanol. This methanol contaminant can be harmful to humans if consumed in excess, and so it is removed during the distillation process.

Because the boiling points of methanol (65°C) and ethanol (78°C) are similar, samples of distillate are tested to ensure most of the methanol has been removed. This is done by adding iodine and sodium hydroxide, and observing the formation of an iodoform (CHI_3).

The three-step chemical process is shown here.

$$CH_3CH_2OH \xrightarrow[\text{oxidation}]{I_2/OH^-} CH_3CHO \xrightarrow[\text{substitution}]{I_2/OH^-} CI_3CHO \xrightarrow[\text{hydrolysis}]{OH^-} CHI_3 + HCOO^-$$

Question 33 continues on page 105

c   ☺☺☺ Explain how this test can be used to determine if methanol has been sufficiently separated from the alcohol you produced.

3 marks

_____

_____

_____

_____

_____

Question 34 (3 marks) ☺☺☺

The relationship between boiling points and number of carbon atoms is shown in the graph for alkanes, amines and carboxylic acids.

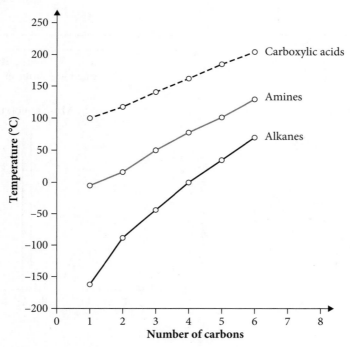

Explain the trends presented in the graph.

_____

_____

_____

_____

_____

_____

_____

_____

_____

End of Question 34

Question 35 (6 marks) ●●●

0.502 g of an organic compound containing carbon, hydrogen and oxygen was combusted in oxygen to produce carbon dioxide and water. The mass of water produced was 0.300 616 g and the volume of carbon dioxide measured at 25°C and 100 kPa was 0.414 22 L.

Spectroscopic data for the compound is provided. Both ^{13}C and ^{1}H NMR spectra showed two singlets, and no splitting pattern was observed in the ^{1}H NMR spectrum.

Determine the molecular formula and the structural formula of the organic compound with reference to the various data provided.

^{13}C NMR spectrum

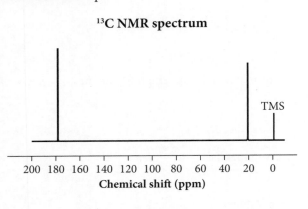

^{1}H NMR spectrum

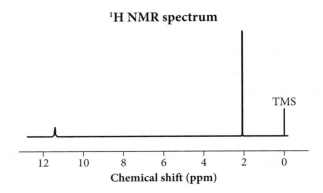

Infrared spectrum

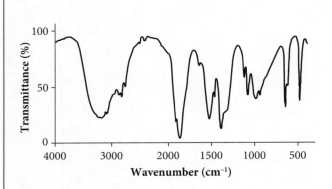

Mass spectrum

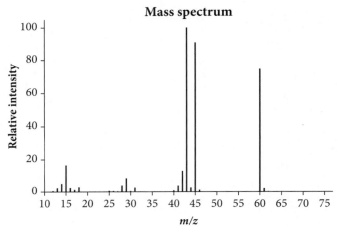

Question 35 continues on page 107

END OF PAPER

9780170465274

SECTION II EXTRA WORKING SPACE

Chemistry

PRACTICE HSC EXAM 2

General instructions
- Reading time – 5 minutes
- Working time – 3 hours
- Write using black pen
- Draw diagrams using pencil
- Calculators approved by NESA may be used
- A formulae sheet, data sheet and Periodic Table are provided at the back of this paper

---

Total marks: 100

Section I – 20 marks
- Attempt Questions 1–20
- Allow about 35 minutes for this section

Section II – 80 marks
- Attempt Questions 21–32
- Allow about 2 hours and 25 minutes for this section

Question 1

Chlorine gas is bubbled through a sample of hex-1-ene. Which type of reaction will occur?

A   Addition

B   Substitution

C   Fermentation

D   Chlorination

Question 2

Which of the following equation(s) can be classified as an Arrhenius acid–base reaction?

$$\text{I} \quad HCl(aq) \rightarrow H^+(aq) + Cl^-(aq)$$

$$\text{II} \quad CH_3COOH(aq) + H_2O(l) \rightarrow CH_3COO^-(aq) + H_3O^+(aq)$$

$$\text{III} \quad KOH(aq) \rightarrow K^+(aq) + OH^-(aq)$$

$$\text{IV} \quad H_2PO_4^-(aq) + H_3O^+(aq) \rightarrow H_2O(l) + H_3PO_4(aq)$$

A   I, II, III and IV

B   I only

C   II and III only

D   I and III only

Question 3

Which piece of equipment would be used to deliver a 25.00 mL aliquot in a titration?

A      B      C      D

Question 4

What is the most likely identity of the distillate in the set-up shown?

A   Ethanol

B   Ethanal

C   Ethanoic acid

D   Water

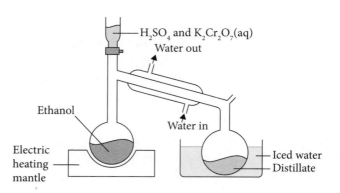

Question 5

Consider the reaction shown.

$$N_2O_4(g) \rightleftharpoons 2NO_2(g)$$

Which statement is true?

A   The mixture becomes a lighter colour when volume is increased.

B   The mixture becomes a darker colour when pressure is increased.

C   The mixture becomes a lighter colour when temperature is decreased.

D   The mixture becomes a darker colour when a catalyst is added.

Question 6

The structure of methyl butanoate is shown.

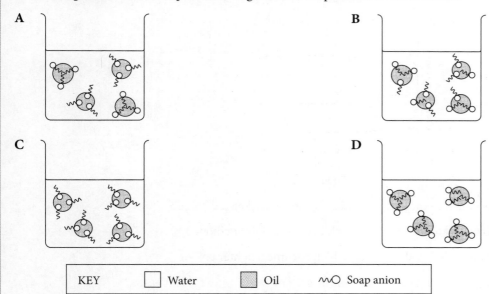

Which pair of chemicals would produce methyl butanoate by esterification?

A   Methanoic acid and butan-1-ol

B   Methanol and butanoic acid

C   Methene and butan-1-ol

D   Methane and butanoic acid

Question 7   ©NESA   2021 SI Q8

Which diagram shows the expected arrangement of soap anions in an emulsion?

A

B

C

D

KEY   ☐ Water   ▨ Oil   ∿O Soap anion

Question 8 ⬤⬤◯

The reaction of aniline ($C_6H_5NH_2$) with water is an equilibrium process.

$$C_6H_5NH_2(l) + H_2O(l) \rightleftharpoons C_6H_5NH^-(aq) + H_3O^+(aq)$$

Identify the acid–base conjugate pair in this reaction.

A $C_6H_5NH^-(aq)$ and $H_2O(l)$

B $C_6H_5NH_2(l)$ and $C_6H_5NH^-(aq)$

C $C_6H_5NH^-(aq)$ and $H_3O^+(aq)$

D $H_3O^+(aq)$ and $C_6H_5NH_2(l)$

Question 9 ⬤⬤◯

What is the conjugate acid of the HPO_4^{2-} ion?

A PO_4^{3-}

B $H_2PO_4^-$

C H_3PO_4

D H_4PO_4

Question 10 ⬤⬤◯

A saturated solution of $Ca(IO_3)_2$ contains $0.0117\,mol\,L^{-1}$ of $Ca(IO_3)_2$. What is the K_{sp} of $Ca(IO_3)_2$?

A 1.37×10^{-4}

B 2.73×10^{-4}

C 6.41×10^{-6}

D 8.42×10^{-10}

Question 11 ⬤⬤◯

Which statement correctly identifies the change at t_1 shown in the graph?

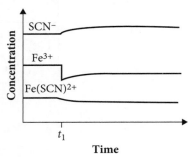

A   Aqueous iron(III) nitrate was added to the system, which became a darker red colour.

B   Aqueous iron(III) nitrate was added to the system, which became a lighter colour.

C   Aqueous sodium hydroxide was added to the system, which became a lighter colour.

D   The pressure was decreased by increasing the volume of the system.

Question 12

What is the correct structure of the condensation polymer formed from these monomers?

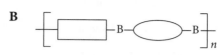

A

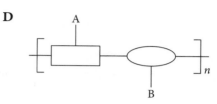

C

B

D

A

B

Question 13

The titration curve below is produced when an acid is titrated with a sodium hydroxide solution of the same concentration.

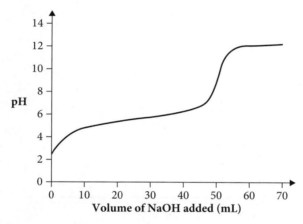

Which indicator would be the most suitable to use in this titration?

A   Bromocresol green (pH range 3.8–5.4)

B   Azolitmin (pH range 4.5–8.3)

C   Cresolphthalein (pH range 8.2–9.8)

D   Indigo carmine (pH range 11.4–13.0)

Question 14 ©NESA 2019 SA Q17

A student makes a solution with a final volume of 200 mL by mixing 100 mL of 0.0500 mol L^{-1} barium nitrate solution with 100 mL of 0.100 mol L^{-1} sodium hydroxide solution.

Which row of the table correctly identifies if a precipitate will form under these conditions, and the reason?

	Will a precipitate form?	Reason
A	Yes	$Q > K_{sp}$
B	Yes	$Q < K_{sp}$
C	No	$Q > K_{sp}$
D	No	$Q < K_{sp}$

Question 15 ⬤⬤

When 10 mL of each of the following $0.1\,mol\,L^{-1}$ solutions are mixed, which pair would produce the highest mass of precipitate?

A   Calcium hydroxide and sodium chloride

B   Calcium hydroxide and copper(II) sulfate

C   Calcium hydroxide and magnesium nitrate

D   Calcium hydroxide and potassium sulfate

Question 16 ⬤⬤⬤

How many hydrogen environments are there in 2-methylpropan-2-ol?

A   2

B   3

C   4

D   7

Question 17 ⬤⬤⬤

A ^{13}C NMR spectrum is shown.

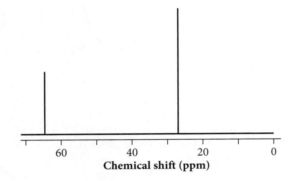

What is the identity of the compound?

A   Ethane

B   Propane

C   Propan-1-ol

D   Propan-2-ol

Question 18 ⬤⬤⬤

Two colourless solutions each contain either Cl^- or CO_3^{2-} anions. Which aqueous reagent could be used to identify both solutions?

A   Lead(II) nitrate

B   Nitric acid

C   Silver nitrate

D   Sodium hydroxide

Question 19 ⬤⬤⬤

The absorbance spectrum of a solution is shown.

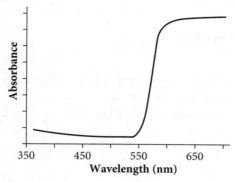

Which ion is the solution likely to contain?

A   Barium

B   Chloride

C   Copper

D   Sodium

Question 20 ⬤⬤⬤

The iron (Fe^{2+}) content of a tablet was determined by first crushing six tablets and then dissolving 2.148 g of the resulting powder in nitric acid. The sample was then oxidised, using potassium permanganate solution, according to the equation:

$$MnO_4^-(aq) + 8H^+(aq) + 5Fe^{2+}(aq) \rightarrow Mn^{2+}(aq) + 5Fe^{3+}(aq) + 4H_2O(l)$$

The resulting solution was made up to a volume of 1.00 L, and then 25.00 mL of this solution was diluted to 250.0 mL. The absorbance of this solution was 0.28 at 525 nm. The calibration curve for MnO_4^- is shown. (1 mol = 1000 mmol)

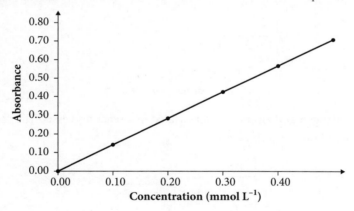

What was the iron (Fe^{2+}) mass in one tablet?

A   0.556 g

B   0.931 g

C   55.6 mg

D   93.1 mg

Section II

80 marks
Attempt Questions 21–32
Allow about 2 hours and 25 minutes for this section

Instructions
- Answer the questions in the spaces provided. These spaces provide guidance for the expected length of response.
- Show all relevant working in questions involving calculations.
- Extra writing space is provided at the back of this booklet. If you use this space, clearly indicate which question you are answering.

Question 21 (9 marks)

Ammonia is manufactured industrially by the Haber process. The chemical reaction is shown below.

$$N_2(g) + 3H_2(g) \rightleftharpoons 2NH_3(g) \qquad \Delta H = -92 \, \text{kJ mol}^{-1}$$

At 400°C the equilibrium constant of this reaction is 1.60×10^{-4} and the activation energy of the forward reaction is approximately $4.00 \times 10^2 \, \text{kJ mol}^{-1} \text{L}^{-1}$.

a   Write the equilibrium expression for this reaction.   1 mark

---

b   Draw the energy profile diagram for this process, labelling E_a, ΔH and products.   3 marks

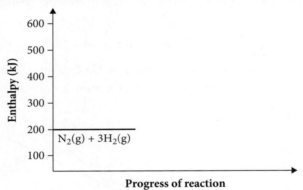

c   Some hydrogen, nitrogen and ammonia were sealed in a reaction vessel. The changes in concentration were recorded on the graph below.

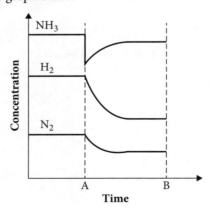

Question 21 continues on page 117

A change was made to the system at time A. Identify this change and use collision theory to explain the shapes of the curves between times A and B.

3 marks

d ●● The temperature of the reaction vessel was increased at time B. On the graph in part **b**, draw how this would affect the concentrations of hydrogen, nitrogen and ammonia as the system returned to equilibrium.

2 marks

Question 22 (9 marks)

Vaping is claimed to be a 'healthier' alternative to smoking. Although it still contains nicotine, the vapour is said to contain fewer chemicals than traditional cigarettes. The nicotine content in each vape cartridge can be determined by titration.

The nicotine is titrated against a solution containing perchloric acid ($HClO_4$) with acetic acid. The equation for this reaction is:

$$HClO_4 + CH_3COOH \rightleftharpoons CH_3COOH_2^+ + ClO_4^-$$

In the titration, the nicotine reacts with the $CH_3COOH_2^+$ ion.

The perchloric acid–acetic acid solution must be standardised with a primary standard before it can be used.

a ●○ Other than having a relatively high molar mass, state two characteristics required of a substance for it to be a useful primary standard.

2 marks

A brand of vape sells capsules containing 7, 14 or 21 mg of nicotine. Because of a malfunction on the production line, a batch of vape capsules was not labelled and the nicotine content was unknown. A chemist was given the task of identifying the nicotine content.

The chemist extracted all the fluid from 25 vape capsules. The solution was then made up to a total of 100.0 mL with a suitable solvent. 20 mL aliquots of the resulting solution were then titrated with standardised 0.0583 mol L^{-1} perchloric acid–acetic acid solution. An average of 15.3 mL was required to reach the end point.

Question 22 continues on page 118

b   ▢■■ Complete the following table by writing the name of the most suitable piece of equipment
to use for each step of the method.   2 marks

Method step	Required equipment
Measuring exactly 100.0 mL of nicotine-containing solution	
Measuring a 20.0 mL aliquot of the nicotine-containing solution	
Adding the perchloric acid–acetic acid solution to the nicotine-containing solution during the titration	

c   ■■■ Use the chemist's titration data to identify the nicotine content of the vape capsules.
Show all working.

The molecular formula of nicotine is $C_{10}H_{14}N_2$ and the titration reaction is:

$$C_{10}H_{14}N_2 + 2CH_3COOH_2^+ \rightarrow C_{10}H_{16}N_2^{2+} + 2CH_3COOH$$ 5 marks

Question 23 (8 marks)

Ethanol can be manufactured by either of two methods:

- fermentation of sugar, such as sucrose ($C_{12}H_{22}O_{11}$)

- using ethene, which is derived from crude oil.

a   ■■▢ Determine the mass, in grams, of pure ethanol that would be produced by the fermentation
of 1.250 kg of sucrose. The molar mass of sucrose is $342\,g\,mol^{-1}$.   3 marks

Question 23 continues on page 119

b Ethene undergoes an addition reaction to form ethanol. Complete the equation.   1 mark

$$C_2H_4(g) + \boxed{} \xrightarrow{\text{catalyst}} C_2H_5OH(g)$$

c Once manufactured, ethanol has many industrial uses, including the synthesis of propyl ethanoate – a common ester that smells like pears.

Complete the flow chart to illustrate how ethanol becomes propyl ethanoate. Include all reactants and conditions required in the allocated spaces.   4 marks

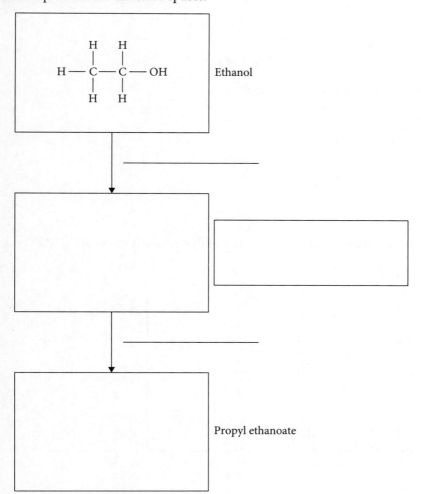

Ethanol

Propyl ethanoate

Question 24 (9 marks)

There are several structural isomers for the molecular formula C_3H_6O.

Three of these are propanal, propanone and prop-2-en-1-ol. The structure of prop-2-en-1-ol is shown below.

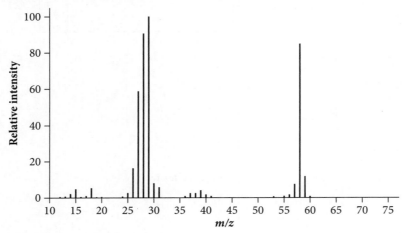

Prop-2-en-1-ol

a   ⬤⬤⬜ Draw the structures of propanal and propanone.   2 marks

b   ⬤⬤⬤ The following mass spectrum was produced by one of these three isomers of C_3H_6O.

i Identify the fragment at *m/z* 29.   1 mark

_____

ii Name the isomer of C_3H_6O that produced this spectrum and justify your choice.   3 marks

_____

_____

_____

_____

_____

Question 24 continues on page 121

c   ☐☐☐ Identify which one of the three named isomers of C_3H_6O produced the following ^{13}C NMR and ^1H NMR spectra. Justify your answer by referring to both spectra.   3 marks

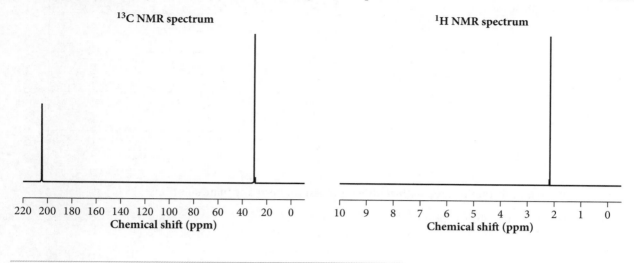

^{13}C NMR spectrum

^1H NMR spectrum

_____

_____

_____

_____

_____

Question 25 (6 marks) ☐☐

The labels have fallen off four reagent bottles containing organic compounds. The substances are:

- cyclohexene

- ethanol

- ethanoic acid

- 2-methylpropan-2-ol.

A student labelled the liquids W, X, Y and Z and conducted tests to identify the liquids. The results are given in the table.

Label	Reaction with acidified potassium permanganate solution	State at room temperature	Miscible with hexane	Miscible with water
W	No	Liquid	Yes	Yes
X	Yes	Liquid	Yes	Yes
Y	No	Solid	Yes	Yes
Z	Yes	Liquid	Yes	No

Question 25 continues on page 122

a   Identify the liquids.   2 marks

Label	Liquid
W	
X	
Y	
Z	

b   For one of the liquids that reacted with acidified potassium permanganate solution:

    i   state an expected observation.   2 marks

_____

_____

    ii   write the IUPAC name and draw the structural formula of a product formed.   2 marks

Question 26 (3 marks) ⬛⬛⬛

Nitrogen dioxide (NO_2) and dinitrogen tetroxide (N_2O_4) exist together in equilibrium.

$$2NO_2(g) \rightleftharpoons N_2O_4(g) \qquad \Delta H = -58 \text{ kJ mol}^{-1}$$
$$\text{brown} \qquad \text{colourless}$$

A gas syringe contains a sample of an equilibrium mixture of the two gases. The mixture is brown in colour.

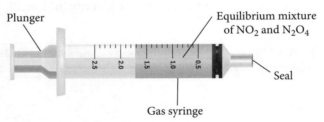

Plunger

Equilibrium mixture of NO_2 and N_2O_4

Seal

Gas syringe

The plunger is pulled outwards to increase the volume. The colour of the mixture first changes to light brown. However, the mixture then turns darker brown as a new equilibrium is established.

State why the mixture changes from light brown to a darker brown. Justify your answer.

_____

_____

_____

_____

End of Question 26

Question 27 (10 marks)

A student set up the following equipment for the fermentation of glucose.

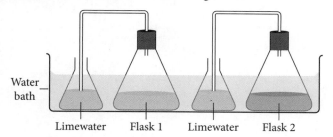

Water bath

Limewater   Flask 1   Limewater   Flask 2

Flask 1 contained 100 mL of a 10% glucose solution.

Flask 2 contained 100 mL of a 10% glucose solution, 2 g of yeast and Pasteur's salts (nutrient for the yeast).

The student recorded the data in a table.

Day	Mass of flask 1 (g)	Mass of flask 2 (g)
0	234.12	235.16
1	234.08	232.05
2	234.00	231.99
3	239.97	230.18
4	239.95	230.18

a Explain the role of flask 1.   1 mark

_____

_____

b Identify the purpose of using the water bath.   2 marks

_____

_____

_____

c Explain the purpose of using the flasks with the limewater.   2 marks

_____

_____

_____

Question 27 continues on page 124

d   ⬤⬤⬤ Calculate the percentage yield of ethanol if the glucose solution contained 67.93 g of glucose. In your response include an equation for fermentation of glucose.

5 marks

Question 28 (5 marks) ⬤⬤⬤

The following equilibrium was established.

$$Fe^{3+}(aq) + SCN^-(aq) \rightleftharpoons Fe(SCN)^{2+}(aq)$$

A student made standard solutions of Fe^{3+} and measured the absorbances. The calibration curve is shown. ($1\,L = 10^6\,\mu L$)

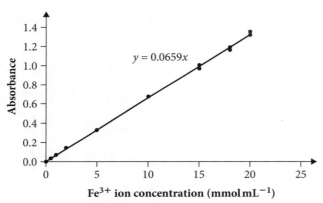

The following data was collected by the student.

Volume of $1.00 \times 10^{-1}\,mol\,L^{-1}$ Fe^{3+} (mL)	Volume of $1.00 \times 10^{-1}\,mol\,L^{-1}$ SCN^- (mL)	Volume of acid solution added (mL)	Absorbance
20.00	20.00	20.00	0.525

Question 28 continues on page 125

Determine the equilibrium constant for this reaction. Show all working.

_____

_____

_____

_____

_____

_____

_____

Question 29 (6 marks)

The following apparatus was used in an experiment to determine the molar enthalpy of combustion of propanol.

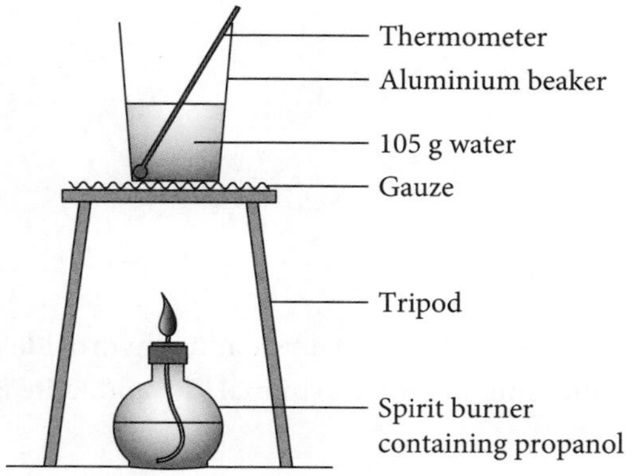

a   Calculate the experimental molar enthalpy of combustion (ΔH) of propanol when 0.632 g propanol was used to increase the water temperature from 16.8°C to 65.0°C.   4 marks

_____

_____

_____

_____

b   The theoretical value for the combustion of propanol is −2644 kJ mol^{-1}. Account for the difference between the experimental value and the theoretical value, referring to the experimental procedure in your response.   2 marks

_____

_____

_____

_____

End of Question 29

Question 30 (7 marks)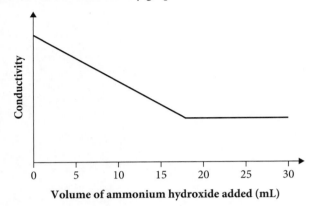

A conductometric titration was undertaken to determine the concentration of an ammonium hydroxide solution. The solution was added to 200.0 mL of a standardised 1.03×10^{-4} mol L^{-1} nitric acid solution. The results of the titration are shown in the conductivity graph.

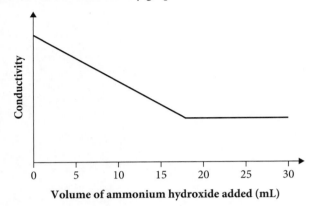

a   Explain the shape of the titration curve.   3 marks

_____

_____

_____

_____

_____

b   The equivalence point was reached when 17.15 mL of ammonium hydroxide had been added. Calculate the concentration of ammonium hydroxide (in mol L^{-1}) and write a relevant chemical equation.   4 marks

_____

_____

_____

_____

_____

_____

Question 31 (4 marks)   ©NESA   2019 SII Q33 (ADAPTED)   ●●●

A student adds 1.17 g of solid aluminium hydroxide to 400 mL of 0.125 mol L^{-1} sulfuric acid.

Calculate the pH of the resulting solution. Assume that the volume of the resulting solution is 400 mL.

_____

_____

_____

_____

Question 31 continues on page 127

_____

Question 32 (4 marks) ©NESA 2020 SII Q23 ●●●

The flow chart summarises an industrial process for the synthesis of ethane-1,2-diol.

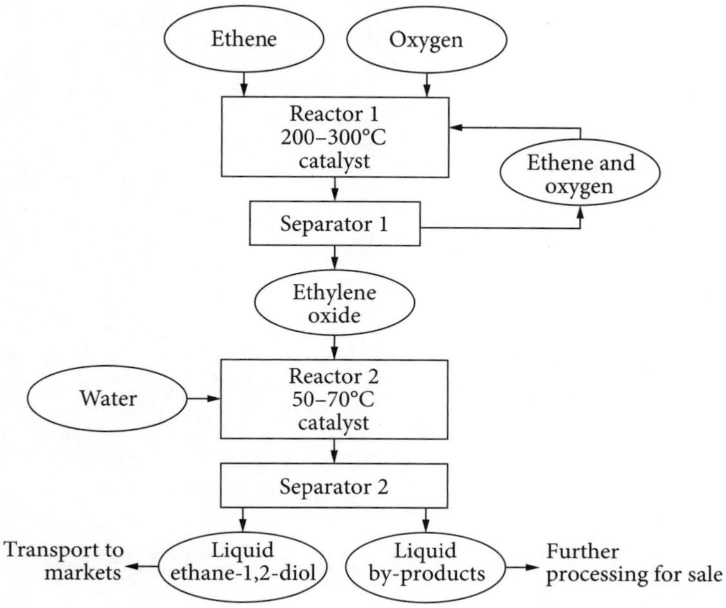

Explain **three** factors that may have been considered in the design of this industrial process. Make specific reference to the flow chart.

_____

END OF PAPER

SECTION II EXTRA WORKING SPACE

SOLUTIONS

Test 1: Static and dynamic equilibrium and factors that affect equilibrium

Multiple-choice solutions

Question 1

A The mass of the reactants equals the mass of the products.

Any reaction will be in dynamic equilibrium if it's reversible and the rates of the forward and reverse reactions are equal. However, equilibrium may favour the production of reactants or products, so the mass of reactants and products may not be equal.

Question 2

B Correct orientation and energy greater than the activation energy

Collision theory states that reactants must collide, and collision energy must be greater than the activation energy for the reaction. Molecules must collide in the correct orientation.

Question 3

D Increasing the concentration of I_2

A is incorrect because increasing pressure favours the side of fewer moles. As both sides of the reaction contain 2 mol, pressure will have no effect. **B** is incorrect because a catalyst increases the rate of reaction but does not change the position of equilibrium. **C** is incorrect because the reaction is exothermic ($\Delta H = -9.4\,\text{kJ mol}^{-1}$); therefore, increasing the temperature will shift the reaction to the endothermic side.

Question 4

C Reaction A, because the activation energy for its reverse reaction is less than that for reaction B

Activation energy is the minimum amount of energy that is required for a successful collision to occur. The lower the activation energy, the less energy that would be required and the more likely a reaction can be reversed.

Question 5

D The rates of both reactions increase, but the rate of the forward reaction increases more than the rate of the reverse reaction.

Increasing temperature increases the rate of reaction of both the forward and reverse reactions because it increases the kinetic energy of the atoms. An increase in temperature will shift the equilibrium in order to remove some of the heat, favouring the endothermic direction.

Question 6

B The concentration of ions increases until equilibrium is established with the water.

The graph increases until it reaches a plateau. The increasing line indicates that the concentration of ions is increasing. The plateau indicates that the concentration of the ions remains constant: this is achieved at equilibrium.

Question 7

B It would become darker brown.

Dinitrogen tetroxide is colourless, whereas nitrogen dioxide is brown. This is an exothermic reaction, so when the temperature increases, the reaction shifts to the endothermic side to mitigate the stress on the system. In this case, it means shifting left to produce more brown nitrogen dioxide.

Question 8

C

Endothermic – this is seen in a profile diagram when the enthalpy of the reactants is lower than the products.

Spontaneous – a reaction that favours the production of products if entropy increases by more than the change in enthalpy.

Question 9

B The concentration of $CoCl_4^{2-}$ would increase.

As a mandated practical within the syllabus, you are expected to know the results, and colours, of these reactions.

A is incorrect because adding NaCl will increase the concentration of Cl^- ions. **C** is incorrect because the solution will be pink at equilibrium. Adding more Cl^- ions will darken the pink colour. **D** is incorrect because adding NaCl will increase the concentration of Cl^- ions.

Question 10

D $4X(g) + 2Y(g) \rightleftharpoons 3Z(g)$ $\quad \Delta H = -100\,kJ$

In all options, as the temperature increases, the yield decreases. This indicates that this is an exothermic reaction because, according to Le Chatelier's principle, increasing temperature in an exothermic reaction will shift the equilibrium to the endothermic side, increasing the concentrations of reactants – this eliminates **A** and **C**.

The yield of Z increases with pressure. According to Le Chatelier's principle, an increase in pressure results in a shift to the side with fewer moles. **B** has 2 mol on both sides and would not be affected by the change in pressure.

9780170465274

Short-answer solutions

Question 11

In an open system, carbon dioxide is free to leave the system, resulting in a decrease in mass of the products compared with in a closed system. According to the law of conservation of mass, the mass of the reactants will equal the mass of the products, which can only be evident in a closed system, where matter cannot be removed.

- 2 marks: links open system to loss of CO_2 and a closed system to containing CO_2 and accounts for the change in mass due to loss
- 1 mark: links open system to loss of CO_2 and a closed system to containing CO_2

Question 12

a $6CO_2(g) + 6H_2O(l) \rightarrow C_6H_{12}O_6(aq) + 6O_2(g)$

b ΔH = products – reactants
$$= (C_6H_{12}O_6(aq) + 6O_2(g)) - (6CO_2(g) + 6H_2O(l))$$
$$= (-1289 + (6 \times 0)) - ((6 \times -381) + (6 \times -287))$$
$$= (-1289) - (-4008)$$
$$= 2719 \, \text{kJ mol}^{-1}$$

ΔS = products – reactants
$$= (C_6H_{12}O_6(aq) + 6O_2(g)) - (6CO_2(g) + 6H_2O(l))$$
$$= (234 + (6 \times 187)) - ((6 \times 236) + (6 \times 54))$$
$$= 1356 - 1740$$
$$= -384 \, \text{J K}^{-1} \text{mol}^{-1} \text{ or } -0.384 \, \text{kJ K}^{-1} \text{mol}^{-1}$$

Numbers for ΔG need to be in the same units. Often, ΔS is converted to kJ.

$\Delta G = \Delta H - T\Delta S$
$$= 2719 - (298 \times -0.384)$$
$$= 2833 \, \text{kJ mol}^{-1}$$

- 5 marks: provides correctly all steps of the equation, including units
- 4 marks: provides correctly all steps of the equation, excluding units
- 3 marks: provides the main steps of the calculation
- 2 marks: provides some relevant steps of the calculation
- 1 mark: provides some relevant information

c Photosynthesis is an orderly ($\Delta S = -384 \, \text{kJ mol}^{-1}$) endothermic reaction ($\Delta H = +2719 \, \text{kJ mol}^{-1}$). The Gibbs free energy of $+2833 \, \text{kJ mol}^{-1}$ indicates that photosynthesis is not a spontaneous reaction. It will only occur when a significant amount of energy is supplied, or a catalyst is used to lower the activation energy. In the presence of chlorophyll, plants convert energy from light into chemical energy in the bonds of the products, absorbing heat from the surroundings as an endothermic reaction.

- 3 marks: uses data to explain ΔH and ΔS of photosynthesis as an orderly endothermic reaction **and** links ΔG to non-spontaneous reaction **and** indicates that a catalyst is required to overcome high activation energy for reaction to proceed
- 2 marks: uses data to explain ΔH and ΔS of photosynthesis as an orderly endothermic reaction **and** links ΔG to non-spontaneous reaction
- 1 mark: provides some relevant information

Question 13

a Some product (NO_2) has been removed from the system.

b

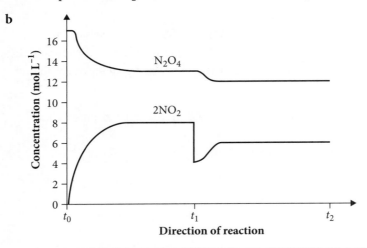

Direction of reaction

> • 2 marks: draws established equilibrium (rise in product, drop in reactant) in a 1 : 2 ratio as per equation
> • 1 mark: draws established equilibrium (rise in product, drop in reactant)

c Decreasing the volume increases the pressure of the system. An increase in pressure increases the rate at which successful collisions occur, increasing the rate of both the forward and reverse reactions. According to Le Chatelier's principle, the reaction shifts to the side of fewer moles, shifting left, to re-establish equilibrium. Therefore, the rate of the reverse reaction is higher until equilibrium.

> • 3 marks: identifies a decrease in volume is an increase in pressure **and** relates an increase in pressure to collision theory (increase of successful collisions and an increase in rate of reaction) **and** links an increase in the rate of the reverse reaction due to Le Chatelier's principle
> • 2 marks: shows a sound understanding of how decreasing volume favours the reverse reaction with reference to some aspects of the collision theory
> • 1 mark: provides some relevant information

Question 14

As heat was added to the system, the rates of the forward and reverse reactions would increase as the increase in energy increases the average kinetic energy of the molecules. This means the fraction of molecules with kinetic energy greater than E_a also increases. Because the energy in the reactants is higher than the energy stored in the bonds of the products, the production of hydrogen chloride gas is exothermic. When the temperature was increased, the system would shift to the left, the endothermic side, to remove some heat from the system, decomposing the hydrogen chloride gas. Therefore, the rate of the reverse reaction would be higher than the rate of the forward reaction, until equilibrium is restored. The reverse reaction can overcome the higher activation energy required due to the extra energy (heat) added to the system.

> • 4 marks: provides an explanation of how an increase in temperature increases the rate of reaction, favouring the endothermic reaction **and** overcomes activation energy **and** refers to energy profile
> • 3 marks: shows understanding of how an increase in temperature increases the rate of reaction, favouring the endothermic reaction
> • 2 marks: shows understanding of what occurs when temperature is increased
> • 1 mark: provides some relevant information

Question 15

a $2NaOH(aq) + H_2CO_3(aq) \rightarrow Na_2CO_3(aq) + 2H_2O(l)$

b Sodium hydroxide reacts with the carbonic acid to form sodium carbonate, removing it from the equilibrium system. In an attempt to restore equilibrium, the reaction drives forward, consuming the available carbon dioxide to produce more carbonic acid. This reaction removes the carbon dioxide from the system faster than just opening the can.

- 3 marks: provides a detailed explanation using Le Chatelier's principle
- 2 marks: shows an understanding of Le Chatelier's principle
- 1 mark: provides some relevant information

Test 2: Calculating the equilibrium constant (K_{eq})

Multiple-choice solutions

Question 1

A All species are in the same state.

Homogenous means 'of the same kind'. When discussing equilibrium, we are looking at the same states.

Question 2

C $\dfrac{[H_2][I_2]}{[HI]^2}$

PORK – products over reactants = K

Molar ratio is moved to the power, e.g. $3X = [X]^3$

Question 3

B Remains unchanged, product

Only temperature changes the value of K_{eq}.

Increasing the concentration of reactants drives the reaction forward to produce more products.

Question 4

D Increase in pressure, decrease in temperature

The increase in pressure favours the side with fewer moles (products); this eliminates **A** and **B**. As an exothermic reaction, a decrease in temperature would drive the reaction forward; this eliminates **C**.

Question 5

A Temperature increased, NO removed

Temperature changes are shown by a gradual change in concentration. Because this is an exothermic reaction, the concentration of product drops when temperature increases. A sharp decrease/increase in the concentration of just one compound indicates the addition of or removal of that particular compound. The changes to the other compounds would have been gradual as they re-established equilibrium.

Question 6

D ii and iii only

(i) There is not enough information provided to determine the rate of the reaction.

(ii) A high K_{eq} lies strongly to the right. So system 2 has more products than reactants.

(iii) Because system 2 has a much smaller K_{eq}, reactants are favoured.

Question 7

A The temperature is decreased.

Temperature is the only factor to change the magnitude of K_{eq}.

Question 8

B 0.22

The reciprocal of 20 is $\dfrac{1}{20} = 0.05$

$\sqrt{0.05} = 0.22$

Question 9

C 6.08×10^{-4}

$0.233 \, \text{mol} = 0.233 \, \text{mol L}^{-1}$ (question states reaction occurs in a 1 L vessel)

	$4FeS_2(s) + 11O_2(g)$	$\rightleftharpoons$	$2Fe_2O_3(s) + 8SO_2$
I	1.00		0
C	$-11x = 0.320$		$+8x$
E	0.68		0.233

$8x = 0.233$

$x = 0.0291$

$K_{eq} = \dfrac{[SO_2]^8}{[O_2]^{11}} = \dfrac{(0.233)^8}{(0.68)^{11}} = 6.08 \times 10^{-4}$

Question 10

D $\dfrac{[B^+][OH^-]}{[BOH]}$

PORK – products over reactants = K

K_b has the same expression as K_{eq}.

Short-answer solutions

Question 11

a

	$2ClF_3(g)$	$3F_2(g)$	$Cl_2(g)$
I	7.5	0.0	0.0
C	$-2x$	$+3x$	$+x$
E	2.5	7.5	2.5

As $x = 2.5$, the change in concentration of ClF_3 is $2.5 \times 2 = 5 \, mol \, L^{-1}$. Therefore, the initial concentration is $5 + 2.5 = 7.5 \, mol \, L^{-1}$.

> • 3 marks: determines the concentration
> • 2 marks: provides a relevant calculation
> • 1 mark: provides some relevant information

b $K_{eq} = \dfrac{[F_2]^3[Cl_2]}{[ClF_3]^2}$

c $K_{eq} = \dfrac{[F_2]^3[Cl_2]}{[ClF_3]^2} = \dfrac{(7.5)^3(2.5)}{(2.5)^2} = 168.75$

> • 2 marks: correctly calculates K_{eq}
> • 1 mark: provides a correct expression

d $K_{eq} = \dfrac{[ClF_3]^2}{[F_2]^3[Cl_2]}$

e $K_{eq} = \dfrac{1}{168.75} = 0.00593$

> • 2 marks: correctly calculates K_{eq} using value from part **c**
> • 1 mark: provides inverse without numerical values

Question 12

a $[CH_4] = \dfrac{c}{V} = \dfrac{1.00}{0.2} = 5 \, mol \, L^{-1}$

$[H_2S] = \dfrac{c}{V} = \dfrac{2.00}{0.2} = 10 \, mol \, L^{-1}$

$[CS_2] = \dfrac{c}{V} = \dfrac{1.00}{0.2} = 5 \, mol \, L^{-1}$

$[H_2] = \dfrac{c}{V} = \dfrac{2.00}{0.2} = 10 \, mol \, L^{-1}$

$Q_{eq} = \dfrac{[CS_2][H_2]^4}{[CH_4][H_2S]^2} = \dfrac{5 \times 10^4}{5 \times 10^2} = 100$

As $Q > K$, the reaction will shift to the left to achieve equilibrium.

> • 3 marks: correctly calculates Q_{eq} **and** compares it with K_{eq} to determine reaction direction
> • 2 marks: provides mostly correct steps to calculate Q_{eq}
> • 1 mark: provides some relevant information

b

$CH_4(g) + 2H_2S(g) \rightleftharpoons CS_2(g) + 4H_2(g)$			
I	5.0	10.0	5.0	10.0
C	$5.00 + x$	$10.0 + 2x$	$5.00 - x$	$10.0 - 4x$
E	6.14	12.28	3.86	5.44

If $[CH_4] = 6.14 \, mol \, L^{-1}$, then $5.00 - x = 6.14$; therefore, $x = 6.14 - 5.00 = 1.14 \, mol \, L^{-1}$

$$Q_{eq} = \frac{[CS_2][H_2]^4}{[CH_4][H_2S]^2} = \frac{3.86 \times 5.44^4}{6.14 \times 12.28^2} = 3.65$$

Equilibrium is reached as $K_{eq} = 3.65$.

- 4 marks: proves calculation to 3 significant figures
- 3 marks: determines correctly the concentration of all compounds at equilibrium
- 2 marks: demonstrates an understanding of how to prove K_{eq}
- 1 mark: provides some relevant information

Question 13

As the temperature increases, the concentration of H_2 increases. This indicates that the reaction is endothermic because more hydrogen fluoride decomposes to counteract the effect of increasing the temperature as it re-establishes a new equilibrium. As the product concentration is now higher than reactant concentration, the Q value would be higher than the original K_{eq}.

- 4 marks: identifies reaction as endothermic **and** determines that K_{eq} value will increase **and** justifies prediction using data from the graph
- 3 marks: identifies reaction as endothermic **and** determines that K_{eq} value will increase
- 2 marks: recognises an increase in $[H_2]$ **and** acknowledges that temperature changes a K_{eq} value
- 1 mark: provides some relevant information

Question 14

a $CO(g) + 2H_2(g) \rightleftharpoons CH_3OH(g)$

$$K_{eq_1} = \frac{[CH_3OH]}{[CO][H_2]^2} = \frac{0.33}{(0.7)(0.37)^2} = 3.44$$

$$K_{eq_2} = \frac{[CH_3OH]}{[CO][H_2]^2} = \frac{0.45}{(0.52)(0.65)^2} = 2.05$$

As the K_{eq} value decreased, the temperature of the system must have increased, shifting the reaction to the left in order to absorb the heat, permanently changing the equilibrium position.

- 4 marks: compares K_{eq} values to determine change to the system
- 3 marks: correctly calculates the equilibrium positions
- 2 marks: provides some steps to calculate an equilibrium position
- 1 mark: provides some relevant information

b   There will be no change to the equilibrium position because only temperature changes a K_{eq} value.

Test 3: Solution equilibria

Multiple-choice solutions

Question 1

B $[Ba^{2+}][NO_3^-]^2$

Question 2

B Lead(II) chloride

Question 3

C $MgCO_3$

Question 4

D $K_{sp} = [Zn^{2+}][OH^-]^2$

Question 5

C 3.6×10^{-8}

Question 6

C Potassium sulfate

Question 7

A Yes, $Q > K_{sp}$

A ppt will form when $Q > K_{sp}$ and will not form when $Q < K_{sp}$; this eliminates **B** and **C**.

$$CuSO_4(aq) + 2AgNO_3(aq) \rightarrow Cu(NO_3)_2(aq) + Ag_2SO_4(s)$$

$$\begin{aligned} n(CuSO_4) &= c \times V \\ &= 0.2 \times 0.150 \\ &= 0.03 \, mol \end{aligned}$$

Ionic ratio 1 : 1; therefore, $[SO_4^{2-}] = 0.03 \, mol \, L^{-1}$

$$\begin{aligned} n(AgNO_3) &= c \times V \\ &= 0.3 \times 0.150 \\ &= 0.045 \, mol \end{aligned}$$

Ionic ratio 1 : 1; therefore, $[Ag^+] = 0.045 \, mol \, L^{-1}$

$$Ag_2SO_4 \rightleftharpoons 2Ag^+ + SO_4^{2-} \quad K_{sp} = 1.20 \times 10^{-5} \text{ (from the data sheet)}$$

$$\begin{aligned} Q &= [0.0045]^2[0.003] \\ &= 6.1 \times 10^{-5} \end{aligned}$$

$Q > K_{sp}$; therefore, ppt will form.

Question 8

A The solubility of barium hydroxide decreases.

Question 9

B 7.44×10^{-6}

$CaCO_3 \rightleftharpoons Ca^+ + CO_3^{2-}$ ionic ratio 1 : 1; therefore, $[Ca^+] = 0.153 \, mol \, L^{-1}$ and $[CO_3^{2-}] = 0.153 \, mol \, L^{-1}$

$$Ag_2CO_3 \rightleftharpoons 2Ag^+ + CO_3^{2-}$$

$$[Ag^+] = x$$

$$K_{sp} = [Ag^+]^2[CO_3^{2-}]$$

$$8.46 \times 10^{-12} = x^2 \times 0.153$$

$$x^2 = \frac{8.46 \times 10^{-12}}{0.153}$$

$$x = \sqrt{5.53 \times 10^{-11}}$$

$$x = 7.44 \times 10^{-6}$$

Question 10

C 1.1×10^{-12}

Short-answer solutions

Question 11

a

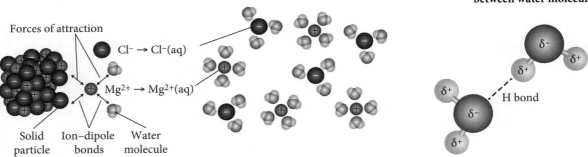

Hydrogen bonding between water molecules

As a solid, magnesium chloride is in an ionic lattice. As a polar molecule, water forms hydrogen bonds with other water molecules. The positive pole of water is attracted through ion–dipole bonds to the negative chloride ions, whereas the negative pole of water is attracted to the positive magnesium ions. The entropy of the system is increased as the ordered lattice of the salt is broken, and the ions are dispersed randomly throughout the solution.

- 4 marks: provides a description of changes in both bonding and entropy **and** supports answer with a labelled diagram
- 2–3 marks: demonstrates some understanding of the changes that occur in bonding **and/or** entropy
- 1 mark: demonstrates an understanding of bonding or entropy **or** draws a labelled diagram

b

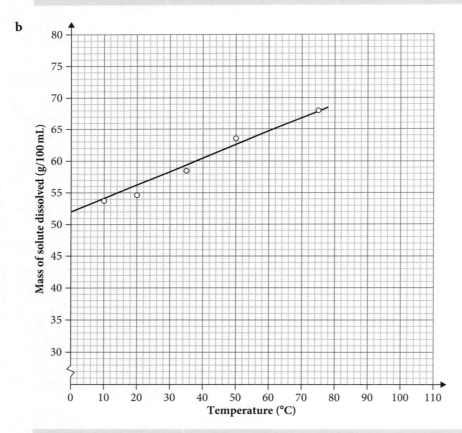

- 3 marks: provides graph with correctly labelled axes, including units, appropriate scales, correctly plotted points, correct line of best fit
- 2 marks: provides a substantially correct graph
- 1 mark: provides a graph with some correct features

c 57–57.5 g

d The student will see that most of the magnesium chloride (121 g) will dissolve, but the remaining 4 g will settle to the bottom and be visible as a solid.

- 2 marks: explains observations
- 1 mark: indicates it will not dissolve

Question 12

a

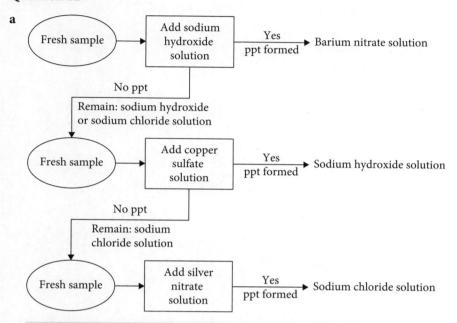

- 3 marks: constructs a flow chart that includes fresh sample each time, clear directions with labels and arrows, correct reagent and observation of chemical
- 2 marks: provides a substantially correct flow chart
- 1 mark: provides a flow chart with some correct features

b $Ba^{2+}(aq) + 2OH^-(aq) \rightarrow Ba(OH)_2(s)$

- 2 marks: provides a correct, balanced net ionic equation
- 1 mark: provides a partially correct net ionic equation

Question 13

a
$$K_{sp} = [Pb^{2+}][Cl^-]^2$$
$$2.4 \times 10^{-4} = [Pb^{2+}] \times 0.025^2$$
$$Pb^{2+} = \frac{2.4 \times 10^{-4}}{0.025^2} = 0.38 \, mol \, L^{-1}$$

- 2 marks: correctly calculates the concentration of lead ions
- 1 mark: provides a relevant step towards the calculation of lead ions

b $[Pb^{2+}] = \dfrac{0.1 \times 0.03}{0.5} = 0.006 \, mol \, L^{-1}$

$[Cl^-] = \dfrac{0.4 \times 0.09}{0.5} = 0.072 \, mol \, L^{-1}$

$Q_{sp} = [Pb^{2+}][Cl^-]^2 = 0.006 \times 0.072^2 = 3.1 \times 10^{-5}$

$K_{sp} > Q_{sp} = 2.4 \times 10^{-4} > 3.1 \times 10^{-5}$; therefore, a precipitate will not form.

- 4 marks: correctly calculates Q_{sp} and makes a judgement statement against K_{sp}
- 3 marks: correctly calculates Q_{sp}
- 2 marks: calculates the number of moles of ions in solution
- 1 mark: provides some relevant working out

c $PbCl_2 \rightleftharpoons Pb^{2+} + 2Cl^-$

$1.94 \, g/100 \, mL \times 10 = 19.4 \, g \, L^{-1}$

$n(PbCl_2) = \dfrac{m}{MM} = \dfrac{19.4}{278.1} = 0.069\,759 \, mol \, L^{-1}$

ion ratio = $[Pb^{2+}] : [Cl^-] = 1 : 2$; therefore, $[Pb^{2+}] = 0.069\,759 \, mol \, L^{-1}$, $[Cl^-] = 0.135\,918 \, mol \, L^{-1}$

$K_{sp} = [Pb^{2+}][Cl^-]^2 = 0.069\,759 \times 0.135\,918^2 = 1.36 \times 10^{-3}$

- 3 marks: correctly calculates K_{sp}
- 2 marks: correctly calculates the number of moles of $PbCl_2$
- 1 mark: provides some relevant working out

Test 4: Properties of acids and bases

Multiple-choice solutions

Question 1

C Antacid, dishwashing detergent, sea water

A is incorrect because urine contains uric acid. **B** is incorrect because orange juice contains citric acid. **D** is incorrect because lemonade contains carbonic acid/citric acid and tomato juice contains citric acid.

Question 2

B Nitric acid and sulfuric acid

Monoprotic acids can donate one proton, whereas diprotic acids can donate two protons. Knowing the chemical formulae of acids and bases is important.

Question 3

B change colour at specific pH values.

An indicator gives a visual sign through colour change of the concentration of hydrogen ions in a solution.

Question 4

C Green to yellow to red

The solution begins as a base (green) and as acid is added, the base is neutralised (yellow) until the acid is in excess, making the solution acidic (red).

Question 5

D sea water is 1000 times higher than in rain water.

The pH scale is a log scale. Each number on the scale is 10× higher than the number below it (remember that a low pH means a higher concentration of H^+ ions).

Question 6

A Yellow, green, colourless

As the pH is 7.3, match the colour that would be seen by each indicator.

Question 7

A $NH_3(aq) + CH_3OH \rightleftharpoons CH_3O^-(aq) + NH_4^+(aq)$, Brønsted–Lowry

A Brønsted–Lowry acid is defined as a proton donor and a base as a proton acceptor. Arrhenius defines an acid as a chemical that ionises in water to produce hydrogen ions and a base as producing hydroxide ions. NH_3 is accepting a proton from CH_3OH to form NH_4^+ (Brønsted–Lowry base). **B** is incorrect because $H^+(aq) + OH^-$ (Arrhenius not Brønsted–Lowry). **C** is incorrect because H_2O is accepting a proton from HNO_3 to produce a hydronium ion (Brønsted–Lowry acid, not Arrhenius). **D** is incorrect because it is not an acid–base reaction.

Question 8

D Hydrochloric acid

Lavoisier defined acids as oxygen-containing species. HCl does not contain oxygen.

Question 9

B $-56.85 \, kJ \, mol^{-1}$

$$q = mC\Delta T = 200 \times 4.18 \times 3.4 = -2842.4 \, J$$
$$NaOH(aq) + HCl(aq) \rightleftharpoons NaCl(aq) + H_2O(l)$$

1:1:1 ratio

$$n(NaOH) = c \times V$$
$$= 0.5 \times 0.100$$
$$= 0.05 \, mol; \text{ therefore, } n(H_2O) = 0.05 \, mol$$
$$\Delta H = -\frac{2842.4}{0.05}$$
$$= -56\,848 \, J \, mol^{-1} \text{ or } -56.85 \, kJ \, mol^{-1}$$

Note: The neutralisation of a strong acid with a strong base always equals approx. $-57 \, kJ \, mol^{-1}$.

Question 10

B 6.72 g

$$NaHCO_3(s) + HNO_3(aq) \rightarrow NaNO_3(aq) + CO_2(aq) + H_2O(l)$$
$$n(HNO_3) = c \times V = 0.4 \times 0.2 = 0.08 \, mol$$

Ratio 1:1; therefore, $n(NaHCO_3) = 0.08 \, mol$

$$m(NaHCO_3) = n \times MM = 0.08 \times 84.007 = 6.72 \, g$$

Short-answer solutions

Question 11

a   For example, for red roses:

1. Put 5 g of red rose petals into a saucepan containing 100 mL of water.

2. Boil the water containing the petals for a few minutes.

3. Filter the petals from the liquid and cool the liquid to room temperature.

> - 2 marks: provides a valid method with numbered steps, logical sequential order, specific quantities and equipment named
> - 1 mark: provides a valid method with numbered steps and logical, sequential order

b   The question provides the pH range of turmeric as 7.4–8.6.

The use of turmeric to classify hydrochloric acid and vinegar as acidic is valid because both have a pH less than 7.4, as indicated by the yellow colour. However, it is unsuitable for differentiating between a strong and a weak acid.

Turmeric remained yellow when added to water and baking soda; therefore, the student incorrectly classified these substances as acidic rather than neutral and weakly basic respectively.

Turmeric allowed the student to correctly classify sodium hydroxide, as seen by the red colour when pH exceeded 8.6.

> - 2 marks: discusses the suitability of the indicator to classify strong acids and bases, weak acids and bases and neutral water
> - 1 mark: discusses the suitability of the indicator to classify strong acids and bases

Question 12

a $H_2SO_4(aq) + 2NaOH(aq) \rightleftharpoons Na_2SO_4(aq) + 2H_2O(l)$

$n(H_2SO_4) = c \times V = 0.10 \times 0.130 = 0.0130\,mol$

$n(NaOH) = c \times V = 0.2 \times 0.065 = 0.0130\,mol$

Molar ratio $H_2SO_4 : NaOH = 1 : 2$; therefore, NaOH is limiting and $n(H_2SO_4) = 0.0065\,mol$

Molar ratio $NaOH : H_2O = 1 : 1$; therefore, $n(H_2O) = 0.0130\,mol$

$q = mC\Delta T$
$= 195 \times 4.18 \times (24.3 - 19.2)$
$= 4157.01\,J$

$\Delta H = -\dfrac{q}{n} = -\dfrac{4157.01}{0.0130} = -320\,kJ\,mol^{-1}$

> - 4 marks: provides balanced chemical equation **and** correctly calculates q **and** correctly calculates number of moles of H_2O **and** calculates ΔH to 3 significant figures
> - 3 marks: provides balanced chemical equation **and** correctly calculates q **and** correctly calculates number of moles of H_2O
> - 2 marks: provides balanced chemical equation **and/or** calculates q **and** attempts to calculate number of moles of H_2O
> - 1 mark: provides some relevant working out

b   Limiting reagent is NaOH.

Excess $n(H_2SO_4) = 0.0130 - 0.0065 = 0.00650\,mol$

> - 2 marks: identifies limiting reagent **and** correctly calculates number of moles
> - 1 mark: identifies limiting reagent **or** correctly calculates number of unreacted moles

c The specific heat capacity of the solution is the same as for water.

All the energy released by the reaction is contained by the Styrofoam cups.

> • 2 marks: two correct assumptions
> • 1 mark: one correct assumption

Question 13

A Brønsted–Lowry acid is a compound that breaks down to donate a proton. In methanol, HCl donates a proton to the alcohol: $HCl(g) + CH_3OH(aq) \rightleftharpoons CH_3OH_2^+(aq) + Cl^-(aq)$.

An Arrhenius acid is a compound that dissociates in water to form a hydrogen ion. Because the HCl is being dissolved in methanol and not water, it cannot be classified as an Arrhenius acid.

> • 3 marks: defines a Brønsted–Lowry and an Arrhenius acid **and** links the dissolution of HCl in different solvents to the definition of acids **and** includes a chemical equation
> • 2 marks: defines a Brønsted–Lowry **and** an Arrhenius acid
> • 1 mark: defines a Brønsted–Lowry **or** an Arrhenius acid

Question 14

a As carbon dioxide dissolves in water, it reacts with the water to form carbonic acid according to the equation:

$$CO_2(g) + H_2O(l) \rightleftharpoons H_2CO_3(aq)$$

When the diprotic carbonic acid dissociates in the water, there is an increase in the concentration of H_3O^+ ions. This reduces the pH of the ocean water.

$$H_2CO_3(aq) + H_2O(l) \rightleftharpoons H_3O^+(aq) + HCO_3^-(aq)$$

$$HCO_3^-(aq) + H_2O(l) \rightleftharpoons H_3O^+(aq) + CO_3^{2-}(aq)$$

> • 4 marks: identifies the formation of carbonic acid **and** links the dissociation of the diprotic acid to decreased pH **and** includes at least two balanced chemical equations
> • 3 marks: identifies the formation of acid **and** links the dissociation of the diprotic acid to decreased pH **and** includes a balanced chemical equation
> • 2 marks: identifies the formation of acid **and** links the dissociation of the acid to decreased pH
> • 1 mark: identifies the formation of acid **or** links the dissociation of the acid to decreased pH **or** includes a balanced chemical equation

b Carbonic acid is reacting with the calcium carbonate in the shells – dissolving them (acid–base reaction). Without shells, these animals are more vulnerable to predators.

Another way to understand this reaction is to view it as an equilibrium reaction and apply your understanding of Le Chatelier's principle.

Reaction 1: $H_2CO_3(aq) + H_2O(l) \rightleftharpoons H_3O^+(aq) + HCO_3^-(aq)$

Reaction 2: $HCO_3^-(aq) + H_2O(l) \rightleftharpoons H_3O^+(aq) + CO_3^{2-}(aq)$

According to the reactions above, calcium carbonate is adding ions (Ca^{2+} and CO_3^{2-}) to the solution. By increasing the concentration of carbonate ions, there is a resultant shift in reaction 2 to the left. This tries to restore equilibrium. This in turn shifts reaction 1 to the left, increasing the concentration of carbonic acid (and making the ocean more acidic).

> • 1 mark: states a valid reason

Test 5: Using Brønsted–Lowry theory

Multiple-choice solutions

Question 1

A $[H^+] = [OH^-]$

In pure water, concentrations of hydrogen ions and hydroxide ions are always equal, according to the equation: $H_2O \rightleftharpoons H^+ + OH^-$. This is why pure water always has a neutral pH. Yet a neutral pH does not need to mean 7!

Question 2

B have large K_a values.

A is incorrect because concentration does not imply strength. You can have high concentrations of both strong and weak acids. **C** is incorrect because pK_a is the negative log of K_a; therefore, a high pK_a is a weak acid. **D** is incorrect because a strong acid completely ionises in solution.

Question 3

C Nitric acid and acetic acid

A – strong acid, strong base. **B** – strong acid, weak base. **D** – weak acid, weak acid

Question 4

C Ethanoic acid, carbonic acid

The larger the K_a value, the stronger the acid. Strong acids have a low pH (a higher concentration of H^+).

Question 5

A HCN

Weak acids produce strong conjugate bases. **B** and **C** are strong acids, producing weak conjugate bases. **A** and **D** are both weak acids, but HCN is weaker than H_3PO_4. Weak acids do not fully ionise in solution. The fact that HCN has only 1 proton and it does not always disassociate implies that CN^- must be a very strong conjugate base to hold onto its proton.

Question 6

B Ionisation increases and pH increases.

As water is added, the number of moles of H_3O^+ remains the same, yet the concentration changes ($c_1V_1 = c_2V_2$).

For example:

$$V_1 = 10.0\,\text{mL} \qquad V_2 = 30.0\,\text{mL} \qquad c_1 = 0.1\,\text{mol L}^{-1} \qquad c_2 = ?$$

$$c_1V_1 = c_2V_2$$

$$c_2 = \frac{c_1 \times V_1}{V_2}$$

$$= \frac{0.1 \times 0.01}{0.03} = 0.03\,\text{mol L}^{-1}$$

$$pH = -\log[H_3O^+] = -\log 0.03 = 1.5$$

Compared with the pH of $0.1 = -\log 0.1 = 1$.

Question 7

D 1.70

$pH = -\log 0.021 = 1.677\,78$, rounding to 1.70

Question 8

B 2.51×10^{-7}

pH of 7.4 = pOH of $14 - 7.4 = 6.6$

$[OH^-] = 1 \times 10^{-6.6} = 2.51 \times 10^{-7}\,mol\,L^{-1}$

Question 9

C 12.90

$pOH = -\log 0.083 = 1.08$

$pH = 14 - 1.08 = 12.9$

Question 10

A

As the concentration of CO_2 increases, the pH should decrease.

A buffer mitigates a change in pH for a short time, before capacity is exceeded and the pH decreases at a greater rate.

Short-answer solutions

Question 11

This is one example of how these terms could be modelled.
This question can be modelled in a series of diagrams.

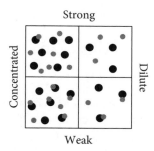

- 3 marks: diagram(s) correctly model all four terms
- 2 marks: diagram(s) correctly model three terms
- 1 mark: diagram(s) correctly model at least one term

Question 12

a The concentration of the solution remains unchanged. $[OH^-] = 0.230\,mol\,L^{-1}$.

b $c_2 = \dfrac{c_1 V_1}{V_2} = \dfrac{0.230 \times 0.020}{0.1} = 0.046\,mol\,L^{-1}$

$[OH^-] = 0.046\,mol\,L^{-1}$; therefore, $pOH = -\log 0.046 = 1.34$

$pH = 14 - 1.34 = 12.66$

- 3 marks: correctly determines the pH of the dilution
- 2 marks: correctly determines the pOH of the dilution
- 1 mark: calculates concentration of dilution

c Adding more water decreases the concentration of OH⁻ ions with a corresponding increase in the concentration of H⁺ ions; therefore, decreasing the pH closer to 7.

> - 2 marks: relates a decrease in OH⁻ ions to a decrease in pH
> - 1 mark: identifies that dilution decreases pH

Question 13

a $pH = -\log 0.01 = 2.00$

b Nitric acid is a strong acid that completely ionises in solution, whereas nitrous acid is a weak acid that does not completely ionise.

The higher degree of ionisation results in a higher concentration of H⁺ ions in the solution.

A higher concentration of H⁺ ions results in a lower pH.

Therefore, nitric acid, with its high degree of ionisation, would have a lower pH than weak nitrous acid.

> - 2 marks: correctly compares the ionisation of both acids **and** identifies that the pH of nitric acid is lower than nitrous acid
> - 1 mark: correctly compares the ionisation of both acids **or** identifies that the pH of nitric acid is lower than nitrous acid

c $n(HNO_3) = c \times V = 0.01 \times 0.05 = 0.0005 \, mol$

$n(Ca(OH)_2) = c \times V = 0.01 \times 0.05 = 0.0005 \, mol$

$$2HNO_3(aq) + Ca(OH)_2(aq) \rightleftharpoons Ca(NO_3)_2(aq) + 2H_2O(l)$$

Therefore, $H^+ : OH^- = 2 : 1$

Nitric acid is limiting; therefore, $n(OH^-)$ remaining $= 0.0005 - 0.00025 = 0.00025 \, mol$

$[OH^-] = \dfrac{n}{V} = \dfrac{0.00025}{0.1} = 0.0025 \, mol \, L^{-1}$

$pOH = -\log 0.0025 = 2.6$

$pH = 14 - 2.6 = 11.40$

> - 4 marks: provides a balanced equation **and** calculates initial moles of HNO₃ and Ca(OH)₂ **and** correctly performs a pH calculation to 2 decimal places
> - 3 marks: completes the calculation with one error
> - 2 marks: provides a balanced equation **and/or** calculates initial moles of HNO₃ and Ca(OH)₂ **and/or** correctly performs a pH calculation
> - 1 mark: provides some relevant information

Question 14

a Amphiprotic ions can donate or accept a proton, so they can act as an acid or a base.

As an acid, hydrogen sulfate donates a proton:

$$HSO_4^- + OH^- \rightleftharpoons SO_4^- + H_2O$$

As a base, hydrogen sulfate accepts a proton from water:

$$HSO_4^- + H^+ \rightleftharpoons H_2SO_4$$

> - 3 marks: correctly defines amphiprotic **and** two chemical equations demonstrating HSO₄⁻ acting as an acid and as a base
> - 2 marks: correctly defines amphiprotic **and** one chemical equation demonstrating HSO₄⁻ acting as an acid and as a base
> - 1 mark: correctly defines amphiprotic **or** one chemical equation demonstrating HSO₄⁻ acting as an acid and a base

b A student could use a probe to determine the pH of the resultant solution.

When hydrogen sulfate ion acts as a base, sulfuric acid is produced; this will be indicated by a low pH.

When hydrogen sulfate ion acts as a base, water and the resultant salt will have a neutral pH.

> - 2 marks: outlines in general terms a valid method for demonstrating the amphiprotic nature of hydrogen sulfate
> - 1 mark: outlines in general terms a valid method for demonstrating hydrogen sulfate as an acid **or** as a base

Question 15

Solution	pH	Justification
Sodium chloride	6.9	$NaCl(aq) + H_2O(l) \rightleftharpoons NaOH(aq) + HCl(aq)$ The ionisation results in a strong conjugate acid and strong conjugate base. The acid and base cancel each other out, resulting in a neutral solution.
Sodium acetate	8.6	$CH_3COONa(aq) + H_2O(l) \rightleftharpoons NaOH(aq) + CH_3COOH(aq)$ The ionisation results in the production of a strong conjugate base and weak conjugate acid. As the weak conjugate acid does not contribute much to the pH, the resulting solution will be basic.
Ammonium nitrate	5.4	$NH_4NO_3(aq) + H_2O(l) \rightleftharpoons NH_4OH(aq) + HNO_3(aq)$ The ionisation results in the production of a strong conjugate acid and weak conjugate base. As the weak conjugate base does not contribute much to the pH, the resulting solution will be acidic.

> - 4 marks: correctly identifies pH of all solutions **and** uses chemical equations to justify
> - 3 marks: correctly identifies pH of all solutions **and** explains how the products affect pH
> - 2 marks: at least two salts correctly matched **and** shows an understanding of products affecting pH
> - 1 mark: identifies one salt

Test 6: Quantitative analysis

Multiple-choice solutions

Question 1

B Measuring cylinder

> A measuring cylinder should not be used to measure the volume of water added. A volumetric flask is calibrated for this purpose.

Question 2

C Acid, base, distilled water

> The burette and pipette should be conditioned with the small volume of the acid or base to ensure all the water has been removed. Excess water will lower the concentration, resulting in an inaccurate calculation. Rinsing the conical flask with water will not affect the number of moles in solution that will react.

Question 3

C Less than $0.200 \, \text{mol} \, \text{L}^{-1}$

It can be solved mathematically without the formula for tartaric acid:

$$H_2A(aq) + 2NaOH(aq) \rightleftharpoons Na_2A(aq) + 2H_2O(aq)$$

$n(NaOH) = c \times V = 0.200 \times 0.0314 = 0.006\,28 \, \text{mol}$

Molar ratio = 1:2; therefore, $n(H_2A) = \dfrac{0.006\,28}{2} = 0.003\,14 \, \text{mol}$

$c(H_2A) = \dfrac{n}{V} = \dfrac{0.003\,14}{0.025} = 0.1256 \, \text{mol} \, \text{L}^{-1}$

Question 4

A Phenolphthalein

Phenolphthalein has a pH range of 8.2–10. Bromothymol blue has a pH range of 6.0–7.3, so **B** is incorrect. Methyl orange has a pH range of 3.0–4.4, so **C** is incorrect. Methyl red has a pH range of 4.4–6.4, so **D** is incorrect.

Question 5

D 10

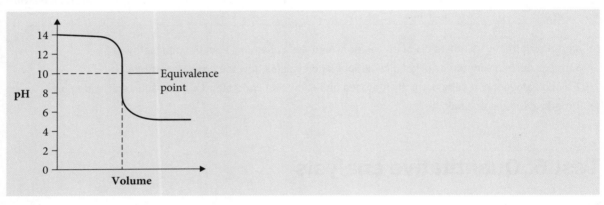

Question 6

B 1.78×10^{-5}

$K_b = 10^{-pK_b}$
$\quad = 10^{-4.75}$
$\quad = 1.78 \times 10^{-5}$

Question 7

A Ineffective, HNO_3 is a strong acid

A buffer consists of a weak acid and its conjugate base. HNO_3 is a strong acid; therefore, this solution would not produce a buffer.

Question 8

D

A strong base completely ionises; therefore, the shape of the curve should begin with high conductivity.

As the strong acid is added, and the base is neutralised, the conductivity decreases until the solution reaches the end point. As the strong acid is added in excess, the high ionisation of the acid increases the conductivity once again.

Question 9

B

> The length of the beginning of the curve and its very slow decline towards the end point suggest the base solution is a buffer. **A** is incorrect because the curve begins at a low pH, indicating a base is being added rather than HCl as per the question. **C** is incorrect because when an acid is added to a base, the pH will decrease, not increase. **D** is incorrect because although a buffer maintains pH until it reaches capacity, the pH will not increase before it decreases with the addition of an acid.

Question 10

C 4.0

> A buffer solution buffers around the pH that is equal to the pK_a of the weak acid that is used to produce the buffer. In this instance, the pH closest to the pK_a is pH 4.0.

Short-answer solutions

Question 11

a Sodium hydroxide is unsuitable as a primary standard because it has a small molecular mass and readily absorbs water from the atmosphere. This will result in a solution that has a concentration lower than calculated, which will give inaccurate results.

An ideal primary standard has a high molecular weight, has high purity and does not react with moisture or other compounds in the air.

> • 3 marks: describes the properties of a standard solution **and** evaluates the use of NaOH as a primary standard
> • 2 marks: describes a property of a standard solution **and** evaluates the use of NaOH as a primary standard
> • 1 mark: describes the properties of a standard solution **or** outlines why NaOH cannot be a primary standard

b

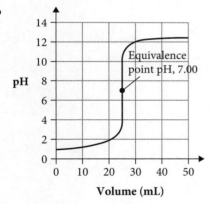

c The most suitable indicator for this titration is phenolphthalein.

The steep neutralisation of the titration curve occurs between pH 4 and pH 10.

Because phenolphthalein has a pH range of 8.2–10 this indicator would change colour to pink at the end point.

> • 2 marks: identifies phenolphthalein as a suitable indicator **and** explains the selection of indicator based on the titration curve
> • 1 mark: identifies phenolphthalein as a suitable indicator

Question 12

$$CH_3COOH(aq) + NaOH(aq) \rightleftharpoons CH_3COONa(aq) + H_2O(aq)$$

$n(NaOH) = c \times V = 0.0145 \times 0.0225 = 0.00032625 \, mol$

Molar ratio = 1 : 1; therefore, $n(CH_3COOH) = 0.00032625 \, mol$

$c(CH_3COOH)diluted = \dfrac{n}{V} = \dfrac{0.00032625}{0.025} = 0.01305 \, mol \, L^{-1}$

$n(CH_3COOH)undiluted = c \times V \times dilution \, factor = 0.01305 \times \dfrac{200}{20} = 0.1305 \, mol$

$m(CH_3COOH) = n \times MM = 0.1305 \times 60.052 = 7.84 \, g$

5% of 8 g = 0.4 g

Therefore, the acceptable mass of ethanoic (acetic) acid is 7.6–8.4 g. The student's calculated mass of 7.84 g confirms the manufacturer's claim.

- 6 marks: provides a balanced chemical equation, including states, **and** uses molar ratio to determine the number of moles of CH_3COOH **and** calculates the concentration of CH_3COOH in the diluted sample **and** calculates mass of CH_3COOH in the undiluted sample **and** determines the acceptable range of CH_3COOH according to manufacturer **and** makes a judgement statement confirming manufacturer's claim
- 5 marks: as above with minor error in calculation
- 4 marks: provides a balanced chemical equation **and** uses molar ratio to determine the number of moles of CH_3COOH **and** calculates the concentration of CH_3COOH in the diluted sample **and** calculates mass of CH_3COOH **and/or** makes a judgement statement about manufacturer's claim
- 3 marks: provides a balanced chemical equation **and** uses molar ratio to determine the number of moles of CH_3COOH **and** calculates the concentration of CH_3COOH in the diluted sample
- 2 marks: provides a balanced chemical equation **and** calculates the number of moles of NaOH
- 1 mark: provides some relevant information

Question 13

a The curve starts with high conductivity due to the highly conductive H^+ ions from the strong acid.

As ammonia is added, H^+ ions are removed, forming ammonium sulfate, steadily decreasing the conductivity until it reaches equivalence at 15 mL.

Conductivity then plateaus as ammonia is added in excess because of the low dissociation of a weak base.

- 3 marks: explains the shape of the graph in terms of the ions present
- 2 marks: describes the trend in the graph
- 1 mark: provides some relevant information

b $H_2SO_4(aq) + 2NH_3(aq) \rightarrow (NH_4)_2SO_4(aq)$

$n(H_2SO_4) = c \times V = 1.12 \times 10^{-4} \times 0.2 = 2.24 \times 10^{-5} \, mol$

Molar ratio is 1 : 2; therefore, $n(NH_3) = 2 \times 2.24 \times 10^{-5} = 4.48 \times 10^{-5} \, mol$

$c(NH_3) = \dfrac{n}{V} = \dfrac{4.48 \times 10^{-5}}{0.015} = 2.99 \times 10^{-3} \, mol \, L^{-1}$

- 4 marks: provides correct calculation and a balanced chemical equation with states
- 3 marks: provides partially correct calculation and a correct balanced equation **or** provides substantially correct calculation
- 2 marks: provides some relevant steps
- 1 mark: provides some relevant information

Question 14

a A buffer is a solution made by mixing a weak acid (citric acid) with its strong conjugate base (sodium citrate) to produce the following buffer system:

$$C_6H_8O_7(aq) \rightleftharpoons C_6H_5O_7^{3-}(aq)$$

This solution can withstand small changes in pH; hence, when small volumes of strong acid or base were added, there was no visible colour change.

$$C_6H_8O_7(aq) + 3OH^-(aq) \rightleftharpoons C_6H_5O_7^{3-}(aq) + 3H_2O(l)$$

$$C_6H_5O_7^{3-}(aq) + 3H^+(aq) \rightleftharpoons C_6H_8O_7(aq)$$

- 4 marks: defines 'buffer' using citric acid and citrate in a chemical equation **and** explains what occurs when an acid and a base are added using relevant chemical equations
- 3 marks: defines 'buffer' **and** explains what occurs when an acid and a base are added **or** includes a chemical equation
- 2 marks: shows some understanding of what occurs when an acid and a base are added
- 1 mark: provides some relevant information

b By increasing the concentration of citric acid and sodium citrate, there will be higher concentrations of $C_6H_8O_7$ and $C_6H_7O_7^-$ in the system. This will allow the buffer to absorb higher volumes of acid or base before a change in pH is witnessed.

- 2 marks: relates an increase in citric acid and sodium citrate to increased volumes of acid and/or base
- 1 mark: provides some relevant information

Test 7: Hydrocarbons and products of reactions involving hydrocarbons

Multiple-choice solutions

Question 1

C Propanone

A is incorrect because it is the common name and not the IUPAC name. **B** is an aldehyde, but the compound shown is a ketone. **D** is incorrect because there is no need to have the number 2 because there is only one possible position of carbonyl group for it to be a ketone.

Question 2

D positional isomers.

The bromine atom is on carbon 2 in X and on carbon 1 in Y. **A** is incorrect because isotopes are of elements with different numbers of neutrons. **B** is incorrect because chain isomers have different chain lengths. **C** is incorrect because X and Y do not have different functional groups.

Question 3

C   Propene + HBr, Propene + HBr

According to Markovnikov's rule, both products, X and Y, will form when propene is reacted with hydrogen bromide, although X will be the major product and Y will be the minor product. **A** and **B** are incorrect because prop-2-ene is the incorrect name and the addition of bromine would produce only one product, 1,2-dibromopropane. **D** is incorrect because the addition of bromine would produce only one product, 1,2-dibromopropane

Question 4

B   *N*-Methylbutanamine

A and **C** are incorrect because there is no amide functional group. **D** is incorrect because the amine is named by using the longest chain.

Question 5

A   Alcohol

Hydration of unsaturated hydrocarbon with a molar mass of $28.052 \, \text{g mol}^{-1}$, ethene forms ethanol, with a molar mass of $46.068 \, \text{g mol}^{-1}$, which has the alcohol functional group.

B, **C** and **D** are all incorrect because they would not yield the desired products.

Question 6

B   Linear, linear, tetrahedral, trigonal planar, trigonal planar

The geometry around each carbon atom in a triple bond (C1 and C2) is linear, whereas the geometry around a carbon atom with four single bonds (C3) is tetrahedral, and around a carbon–carbon double bond (C4 and C5) is trigonal planar.

Question 7

C   Straight, branched

A, **B** and **D** all incorrectly describe the graph.

Question 8

C   It should be handled in a fume cupboard and poured into a waste organic solvent bottle.

A is incorrect because chemicals should never be poured back into the original reagent bottles, to minimise contamination. **B** is incorrect because cyclohexane is not miscible with water, hence should not be poured down the sink. **D** is incorrect because it will not react with baking powder.

Question 9

D I, II and III

Question 10

A 6.946 g

$$n(H_2) = \frac{5.726}{24.79} = 0.230\,98\ldots \text{mol}$$

$$MM(\text{alkene}) = \frac{m(\text{alkene})}{n(\text{alkene})} = \frac{6.473}{0.230\,98\ldots} = 28.024\ldots$$

Therefore, the alkene must be ethene, and the alkane formed must be ethane.

$$m(\text{ethane}) = n \times MM = 0.230\,98\ldots \times ((2 \times 12.01) + (6 \times 1.008)) = 6.946 \text{ g}$$

B is incorrect. It is based on STP, i.e. $n(H_2) = \dfrac{5.726}{22.71} = 0.2521\ldots \text{mol}$

$$m(\text{ethane})\ n \times MM = 0.2521\ldots \times ((2 \times 12.01) + (6 \times 1.008)) = 7.577 \text{ g}$$

C and **D** are incorrect because they are the molar masses of ethene and ethane respectively.

Short-answer solutions

Question 11

a mass (petrol) = density × volume

$m(\text{petrol}) = 0.690 \times 60 \times 1000 = 41\,400 \text{ g}$

$$n(\text{petrol}) = \frac{41\,400}{114} = 363.1578\ldots \text{mol}$$

The energy released, q:

$q = \Delta H \times n = 5460 \times 363.1578\ldots = 1\,986\,473.684 \text{ kJ} = 1.99 \times 10^6 \text{ kJ (to 3 sig. fig.)}$

- 3 marks: provides correct answer **and** correct significant figures **and** correct unit
- 2 marks: provides correct answer **and/or** correct significant figures/unit
- 1 mark: calculates number of moles of petrol correctly

b $n(CH_4) = \dfrac{1\,986\,473.684}{74.85} = 26\,539.39 \text{ mol}$

$V(CH_4) = 24.79 \times n = 24.79 \times 26\,539.39 \text{ L} = 657\,911.5916 \text{ L} = 6.58 \times 10^5 \text{ L (to 3 sig. fig.)}$

- 3 marks: correct answer **and** correct significant figures **and** correct unit
- 2 marks: correct answer **and/or** correct significant figures/unit
- 1 mark: calculates number of moles of methane correctly

Question 12

Compound	IUPAC name	Structural formula
C_4H_8	But-2-ene	(see structure below)
A	2-Bromobutane	(see structure below)
B	Butane	(see structure below)
C	Butan-2-ol	(see structure below)

But-2-ene (C_4H_8):
```
      H   H       H
      |   |       |
  H — C — C = C — C — H
      |       |   |
      H       H   H
```

2-Bromobutane (A):
```
      H   H   H   H
      |   |   |   |
  H — C — C — C — C — H
      |   |   |   |
      H   Br  H   H
```

Butane (B):
```
      H   H   H   H
      |   |   |   |
  H — C — C — C — C — H
      |   |   |   |
      H   H   H   H
```

Butan-2-ol (C):
```
      H   H   H   H
      |   |   |   |
  H — C — C — C — C — H
      |   |   |   |
      H   OH  H   H
```

- 4 marks: states correct IUPAC names for all four compounds **and** draws correct structural formulae
- 3 marks: states correct IUPAC names for three compounds **and** draws correct structural formulae
- 2 marks: states correct IUPAC names for two compounds **and** draws correct structural formulae
- 1 mark: states correct IUPAC names for one compound **and** draws correct structural formula

Question 13

Name	Structural formula
But-1-ene	(see structure below)
But-2-ene	(see structure below)
Cyclobutane	(see structure below)
Methylpropene	(see structure below)

But-1-ene:
```
  H       H   H   H
   \      |   |   |
    C = C — C — C — H
   /          |   |
  H       H   H   H
```

But-2-ene:
```
      H               H
      |               |
  H — C — C = C — C — H
      |   |   |   |
      H   H   H   H
```

Cyclobutane:
```
      H   H
      |   |
  H — C — C — H
      |   |
  H — C — C — H
      |   |
      H   H
```

Methylpropene:
```
          H
          |
      H — C — H
          |
          H
          |
  H — C = C — C — H
      |       |
      H       H
```

- 4 marks: draws four correct structural formulae **and** states correct IUPAC names
- 3 marks: draws three correct structural formulae **and** states correct IUPAC names
- 2 marks: draws two correct structural formulae **and** states correct IUPAC names
- 1 mark: draws one correct structural formula **and** states correct IUPAC name

Question 14

a

```
    Cl          Cl
      \        /
       C  ==  C
      /        \
    Cl          H
```

b The truck is positioned above a metal/reinforced laminated resin with a sloping pipe so that any spills can be naturally drained into the drip pans. The stainless-steel tank is positioned above concrete that is protected by metal/reinforced laminated resin to prevent seepage through the concrete to the soil.

> 'Analyse' questions require you to identify components and the relationship between them; draw out and relate implications.
> - 3 marks: identifies the components **and** relates to safe transfer and storage in a thorough manner
> - 2 marks: identifies the components **and** relates to safe transfer and storage in a limited manner
> - 1 mark: identifies the components **and/or** relates to safe transfer and storage

Question 15

1. Label the bottles 'A' and 'B' and place them in a fume cupboard.

2. Pour 20 mL of each liquid into two labelled beakers.

3. Bubble hydrogen chloride gas through both beakers.

4. Fractionally distil the contents of each beaker.

5. The beaker that contains only one substance that distilled over must have had 3-chlorohexane in it. Hence, the original liquid must have been hex-3-ene.

6. The beaker that resulted in two distillates contains two substances, i.e. must have had 1-chlorohexane and 2-chlorohexane in it. Hence, the original liquid must have been hex-1-ene because Markovnikov's rule predicts two products will be formed.

> - 3 marks: labels the two bottles containing the liquids **and** states quantities used for testing **and** identifies a logical sequence
> - 2 marks: states quantities used for testing **and** identifies a logical sequence
> - 1 mark: identifies a logical sequence

Question 16

Name	Drawing
Pentan-1-ol	![structure: H—C—C—C—C—C—O—H chain with H atoms on each carbon]
Pentan-2-ol	![structure: H—C—C—C—C—C—H chain with O—H group on 2nd carbon from right]
Pentan-3-ol	![structure: H—C—C—C—C—C—H chain with O—H group on middle carbon]

- 4 marks: correctly identifies that alcohols contain the hydroxyl group (–OH) **and** correctly draws positional isomers by changing the position of the hydroxyl group along the chain **and** draws three correct structural formulae **and** states correct IUPAC names
- 3 marks: correctly identifies that alcohols contain the hydroxyl group (–OH) **and** correctly draws positional isomers by changing the position of the hydroxyl group along the chain **and** draws three correct structural formulae **and/or** states correct IUPAC names
- 2 marks: correctly identifies that alcohols contain the hydroxyl group (–OH) **and** correctly draws positional isomers by changing the position of the hydroxyl group along the chain **and** draws one or two correct structural formulae **and/or** states correct IUPAC names
- 1 mark: correctly identifies that alcohols contain the hydroxyl group (–OH) **and/or** draws one isomer **and/or** states correct IUPAC name

Test 8: Alcohols

Multiple-choice solutions

Question 1

B Butan-2-ol

The longest carbon chain has four carbons and the hydroxyl group is on carbon 2. **A** is incorrect because the position of the hydroxyl group is not identified. **C** and **D** are incorrect because they would have 5 and 3 carbons respectively.

Question 2

C 4-Methylpentan-1-ol

The carbon with the hydroxyl group takes priority in numbering. The longest continuous chain is five carbons long and has a methyl group attached to carbon number 4. **A** is incorrect because the longest continuous chain does not contain six carbons. **B** is incorrect because the carbon with the hydroxyl group should have top priority. **D** is incorrect because there is no ethyl group.

Question 3

D Butan-2-ol

Butan-2-ol is an asymmetrical alkanol and will produce but-1-ene and but-2-ene. **A**, **B** and **C** will produce just one alkene – ethene and propene respectively.

Question 4

A Propanal, dilute H_2SO_4

Propan-1-ol is a primary alcohol that will oxidise to the aldehyde propanal by reaction with acidified $K_2Cr_2O_7$. Dilute H_2SO_4 is used to acidify the reaction. **B** is incorrect because concentrated sulfuric acid is not used to acidify the reaction. It is a dehydrating agent. **C** and **D** will be produced from the oxidation of propan-2-ol.

Question 5

C Propan-1-ol. The colour of the $K_2Cr_2O_7$ will change from orange to green.

Propan-1-ol is the independent variable and the colour of $K_2Cr_2O_7$ will change from orange to green as Cr^{3+} ions form.

A, **B** and **D** are incorrect – see explanation above.

Question 6

D Ethanol, HBr, substitution

Ethanol undergoes a substitution reaction with HBr. **A** and **B** are incorrect because ethene is not soluble in water. **C** is incorrect because Br_2 will not react with ethanol.

Question 7

C 12 mL

Equation for fermentation:

$$C_6H_{12}O_6(aq) \xrightarrow{\text{yeast enzymes}} 2C_2H_5OH(aq) + 2CO_2(g)$$

From graph $m(CO_2) = 9.5\,g$

$$n(CO_2) = \frac{m}{MM} = \frac{9.5}{12.01 + (2 \times 16.00)} = 0.215\,860\,031\,mol$$

$n(\text{ethanol}) = n(CO_2) = 0.215\,860\,031\,mol$

$m(\text{ethanol}) = n \times MM = 0.215\,860\,031 \times ((2 \times 12.01) + (6 \times 1.008) + 16.00) = 9.944\,239\,945\,g$

$$V(\text{ethanol}) = \frac{\text{mass}}{\text{density}} = \frac{9.9442...}{0.79} = 12.5876...\,mL = 13\,mL$$

Question 8

D Iodoethane

D is correct because the bond energy for C–I is the lowest at $238\,kJ\,mol^{-1}$, making it the easiest bond to break; hence, it is the most reactive.

A, **B** and **C** are incorrect because they all have higher bond energies as given in the table.

Question 9

D Transesterification

The process shown is transesterification.

A is incorrect because condensation produces a small molecule such as water. **B** is incorrect because esterification requires concentrated sulfuric acid. **C** is incorrect because saponification requires 3 mol of NaOH per triglyceride.

Question 10

C 2-Methylpropan-2-ol

2-Methylpropan-2-ol is a tertiary alcohol and cannot be oxidised. The colour will remain purple.
A, **B** and **D** can all be oxidised and a colour change from purple to very pale pink or colourless will be observed.

Short-answer solutions

Question 11

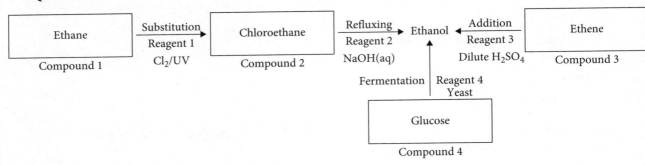

- 3 marks: identifies 4 correct compounds **and** identifies 4 correct reagents
- 2 marks: identifies 2–3 correct compounds **and** identifies 2–3 correct reagents
- 1 mark: identifies 2–4 correct compounds **and/or** identifies 2–4 correct reagents

Question 12

a $C_8H_{18}(l) + 12.5O_2(g) \rightarrow 8CO_2(g) + 9H_2O(g)$

$C_2H_5OH(l) + 3O_2(g) \rightarrow 2CO_2(g) + 3H_2O(g)$

Ethanol requires less oxygen per mole for complete combustion. Octane requires more than four times the amount of oxygen so is more likely to undergo incomplete combustion; hence, there would be more soot at the bottom of the conical flask.

- 3 marks: writes two correct equations for the complete combustion of octane and ethanol **and** compares amount of oxygen required per mole of complete combustion of each fuel **and** recognises that soot is the result of incomplete combustion **and** relates to likelihood of incomplete combustion
- 2 marks: writes two correct equations for complete combustion of octane and ethanol **and/or** compares amount of oxygen required per mole of complete combustion of each fuel **and/or** recognises that soot is the result of incomplete combustion **and/or** relates to likelihood of incomplete combustion
- 1 mark: writes correct equations for complete combustion of octane and/or ethanol **or** provides some relevant information

b

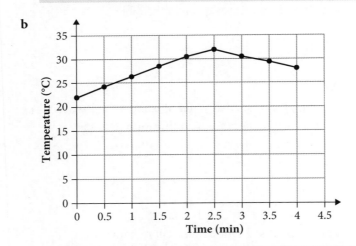

- 3 marks: draws a curve/line of best fit and labels axes – time on x-axis with and temperature on y-axis **and** states unit for each axis **and** an appropriate scale for each axis **and** graph takes up most of the grid
- 2 marks: 3 points from the above
- 1 mark: draws a graph and provides some basic information

c The spirit burner was most likely extinguished after 2 minutes, but before 3 minutes, because the maximum temperature of 32°C was reached at this time. However, it may have reached this temperature just after 2 minutes, or just after 2.5 minutes, but had started to cool by 3 minutes.

- 2 marks: states time range 2 minutes < time < 3 minutes **and** rationale
- 1 mark: states time range 2 minutes < time < 3 minutes **and/or** states 2.5 minutes **and** reason

d ΔT from graph = 32 − 22 = 10°C

m(ethanol burned) = 252.28 − 251.86 = 0.42 g

$$n(\text{ethanol}) = \frac{m}{MM} = \frac{0.42}{(2 \times 12.01) + (6 \times 1.008) + 16.00} = 0.009\,1169\ldots\,\text{mol}$$

Energy released = 0.009 1169… × 745 = 6.792 kJ = 6792 J

$q = mc\Delta T$

$$m(\text{water}) = \frac{q}{c\Delta T} = \frac{6792}{4.18 \times 10} = 162.49\,\text{g}$$

$$V(\text{water}) = \frac{\text{mass}}{\text{density}} = \frac{162.49\,\text{g}}{0.996\,\text{g mL}^{-1}} = 163\,\text{mL (to 3 sig. fig.)}$$

- 3 marks: calculates correct answer **and** correct significant figures **and** correct unit
- 2 marks: correct answer (incorrect significant figures/missing unit)
- 1 mark: calculates correct mass of water

e The experimental value of ΔH of 745 kJ mol^{-1} is less than the theoretical value for the molar heat of combustion of 1367 kJ mol^{-1}. This is due to heat loss and incomplete combustion. The glass conical flask should be replaced with a metal container because it would conduct heat better; the metal container should be insulated and a heat shield should be used.

- 2 marks: states that experimental value is less than the theoretical value **and** identifies two reasons for lower experimental value **and** suggests an improvement
- 1 mark: states that experimental value is less than the theoretical value **and/or** identifies one or two reasons for lower experimental value **and/or** suggests an improvement

Question 13

a 1. Label the bottles X, Y and Z.

2. Label three test tubes X, Y and Z.

3. Place 2 mL of each of X, Y and Z in the appropriate test tube.

4. Add 4–5 drops of acidified potassium permanganate solution to each test tube.

5. Heat the test tubes in a water bath for 1–2 minutes.

6. The solution that stays purple would be 2-methylpropan-2-ol.

7. The solutions that have changed colour from purple to pale pink or colourless need to be fractionally distilled.

8. The mixture that has a fraction distilling at 141°C contains propanoic acid; hence; the original alcohol would have been propan-1-ol.

9. The mixture that has a fraction distilling at 49°C contains propanone; hence; the original alcohol would have been propan-2-ol.

- 3 marks: labels bottles **and** states a logical sequence **and** states observations with oxidising agent **and** refers to different boiling points
- 2 marks: labels bottles **and** states a logical sequence **and** states observations with oxidising agent
- 1 mark: identifies one correct observation with an oxidising agent

b Reduction: $$[MnO_4^-(aq) + 8H^+(aq) + 5e^- \rightarrow Mn^{2+}(aq) + 4H_2O(l)] \times 2$$

Oxidation: $$[CH_3CH(OH)CH_3(l) \rightarrow CH_3COCH_3(l) + 2H^+(aq) + 2e^-] \times 5$$

Net: $$2MnO_4^-(aq) + 6H^+(aq) + 5CH_3CH(OH)CH_3(l) \rightarrow 2Mn^{2+}(aq) + 8H_2O + 5CH_3COCH_3(l)$$

- 3 marks: writes correct equations **and** multiplies by correct quotients for oxidation **and** reduction **and** net equation
- 2 marks: writes correct equations **and/or** multiplies by correct quotients for oxidation **and** reduction
- 1 mark: writes correct equations for oxidation **and/or** reduction

Question 14

Biofuels are produced from renewable sources such as crops rather than from fossil fuels. Biofuels are classified as renewable fuels because they can be continuously produced from crops, algae or animal wastes. The two main biofuels in Australia are ethanol and biodiesel.

Although ethanol is produced mainly from fermentation, it can also be produced industrially through hydration of alkenes. Biodiesel is produced from waste vegetable oil through transesterification reactions.

Engines would need to be modified to use 100% biofuels. This would incur a substantial cost to the economy. Currently ethanol is mixed with petrol in E10, which does not require engine modifications.

Australia could become self-sufficient using its own source of fuel. However, production of bioethanol requires arable land to grow crops, which could otherwise be used to grow food or for shelter. Growing and harvesting crops also requires substantial amounts of energy.

- 3 marks: outlines thoroughly the production of biofuels **and** comments on the viability
- 2 marks: outlines in a limited manner the production of biofuels **and/or** comments on the viability
- 1 mark: outlines in a limited manner the production of biofuels **or** comments on the viability

Test 9: Reactions of organic acids and bases

Multiple-choice solutions

Question 1

B *N*-Ethyl-*N*-methylethanamide

Alphabetical order: therefore, ethyl is written before methyl. There are two alkyl groups, ethyl and methyl, attached to the N.

Question 2

C Methanamine

Litmus turns blue in the presence of a basic substance, and methanamine is basic. **A** and **B** are neutral. **D** is acidic.

Question 3

D Ethanoic acid, propan-2-ol, concentrated sulfuric acid

Production of the ester requires concentrated sulfuric acid.

Question 4

C Separating funnel, no effect on litmus, turns litmus red

The equipment is a separating funnel. Ester is neutral so has no effect on litmus, but the aqueous layer will be acidic because it will contain sulfuric acid from the reaction mixture.

Question 5

B Line B

Alcohols have lower boiling points than carboxylic acids but higher boiling points than ketones and alkanes because of hydrogen bonding. Carboxylic acids have dipole–dipole interactions in addition to hydrogen bonds; therefore, they have higher boiling points than alcohols. Ketones have dipole–dipole interactions and alkanes have dispersion forces only.

Question 6

B a micelle.

The structure is a micelle with the non-polar hydrophobic tails of soap molecules embedded in the grease through dispersion forces, while the ionic head forms ion–dipole forces with the water.

A is incorrect because an emulsion is made up of a number of micelles. **C** is incorrect because a soap ion has an ionic head and a non-polar tail. **D** is incorrect because a grease molecule is likely to be a triglyceride.

Question 7

D As fabric and hair conditioners

The cationic head is attracted to fibres in clothing and hair, so it is used in fabric and hair conditioners.

Question 8

D Saponification

Saponification is the conversion, under alkaline conditions (i.e. in NaOH solution), of fats or oils, such as the triglyceride shown, into glycerol and sodium salts of fatty acids.

Question 9

A Use a cationic detergent because it does not interact with the ions that cause water hardness.

Cationic detergents do not interact with Ca^{2+} and Mg^{2+} ions that cause water hardness.

Question 10

D Propanoic acid is a weaker acid than methanoic acid.

The smaller the pK_a value, the stronger the acid. The pK_a of methanoic acid is less than the pK_a of propanoic acid.

A is incorrect because methanoic acid is the strongest acid. **B** is incorrect because butanoic acid is the weakest acid. **C** is incorrect because ethanoic acid is weaker than methanoic acid.

Short-answer solutions

Question 11

a

The name of the organic product is butyl ethanoate.

> - 3 marks: writes balanced equation using structural formulae **and** uses equilibrium arrows with catalyst written on top **and** correctly names the organic product
> - 2 marks: names the organic product **and** one of the above
> - 1 mark: names the organic product **and/or** one relevant statement

b Esterification is slow at room temperature. Concentrated sulfuric acid is a catalyst that speeds up the rate of reaction by providing an alternative pathway of lower activation energy.

> - 1 mark: identifies that esterification is slow at room temperature **and** concentrated sulfuric acid provides an alternative pathway of lower activation energy

c m(butan-1-ol) = density × volume = 0.81 × 57.80 = 46.818 g

n(butan-1-ol) = $\dfrac{m}{MM}$ = $\dfrac{46.818}{(4 \times 12.01) + (10 \times 1.008) + 16.00}$ = 0.631 65... mol

n(butyl ethanoate) = n(butan-1-ol) = 0.631 65... mol

m(butyl ethanoate) = $n \times MM$
$\qquad\qquad$ = 0.631 65... × ((6 × 12.01) + (12 × 1.008) + (2 × 16.00)) = 73.3700... g

V(butyl ethanoate) = $\dfrac{\text{mass}}{\text{density}}$ = $\dfrac{73.3700...}{0.88}$ = 83.3751... mL

Given 70% yield, V(butyl ethanoate) = 0.7 × 83.3751... mL = 58.36... mL = 58 mL (to 2 sig. fig.)

> - 4 marks: provides correct answer **and** correct significant figures **and** correct unit
> - 3 marks: provides correct answer (incorrect significant figures, missing unit)
> - 2 marks: provides correct mass of butyl ethanoate
> - 1 mark: provides correct number of moles of butan-1-ol

d The reaction mixture can be decanted into a separating funnel to avoid the boiling chips being transferred. Distilled water is added, the mixture is shaken, and the layers are allowed to separate. Then aqueous sodium carbonate is added to the organic layer and the mixture is shaken again. The aqueous layer is then run out. The process is repeated until no more bubbling is observed. The organic layer is fractionally distilled to obtain the pure ester, which should distil at 126°C.

> - 2 marks: describes the separation process thoroughly **and** states removal of aqueous layer **and** repetition until no more bubbling **and** fractional distillation in final step
> - 1 mark: two of the above

Question 12

a

b The pK_a scale is used to measure the strength of acids: $pK_a = -\log_{10}K_a$. So, just like pH, the smaller the pK_a value, the stronger the acid. The pK_a of ethanoic acid is 4.75, making it the strongest acid compared to phenol with pK_a 9.80 and ethanol with pK_a 15.9.

- 2 marks: states what the pK_a scale measures **and** states that the smaller the pK_a, the stronger the acid **and** compares pK_a values of phenol and ethanol
- 1 mark: one of the above

Question 13

a $CH_3CH_2Br + 2NH_3 \rightleftharpoons CH_3CH_2NH_2 + NH_4Br$

- 1 mark: writes a correct net equation

b $n(NH_3) = \dfrac{12.6}{24.79} = 0.508\,269\ldots\ \text{mol}$

The equation shows a $2:1$ ratio of ammonia : ethanamine

$n(\text{ethanamine}) = \dfrac{n(NH_3)}{2} = \dfrac{0.508\,269\ldots}{2} = 0.254\,13\ldots\ \text{mol}$

$V(\text{ethanamine}) = n \times 24.79 = 0.254\,13\ldots \times 24.79 = 6.30\,\text{L (to 3 sig. fig.)}$

- 2 marks: calculates correct volume **and** correct significant figures **and** correct unit
- 1 mark: calculates correct volume (incorrect significant figures/missing unit)

Question 14

a $CH_3CONH_2(l) + H_2O(l) + HCl(aq) \rightarrow CH_3COOH(aq) + NH_4Cl(aq)$

- 1 mark: writes a correct equation with states

b

- 4 marks: writes four correct structural formulae **and** writes three correct reaction conditions **and** must state acidified $K_2Cr_2O_7$ or $KMnO_4$ **and** reflux
- 3 marks: writes three or four correct structural formulae **and** writes two or three correct reaction conditions **and/or** states acidified $K_2Cr_2O_7$ or $KMnO_4$ **and/or** reflux
- 2 marks: writes one or two correct structural formulae **and** writes one or two correct reaction conditions **and/or** states acidified $K_2Cr_2O_7$ or $KMnO_4$ **and/or** reflux
- 1 mark: writes one or two correct structural formulae **and/or** writes one or two correct reaction conditions **and/or** states acidified $K_2Cr_2O_7$ or $KMnO_4$ **and/or** reflux

c V(ethanoic acid) = 64 × 0.45 = 28.8 mL

m(ethanoic acid) = density × V = 1.05 × 28.8 = 30.24 g

n(ethanoic acid) = $\dfrac{m}{MM}$ = $\dfrac{30.24}{(2 \times 12.01) + (4 \times 1.008) + (2 \times 16.00)}$ = 0.503 56… mol

n(ethanamide) = n(ethanoic acid) = 0.503 56… mol

m(ethanamide) = $n \times MM$ = 0.503 56… × ((2 × 12.01) + (5 × 1.008) + 14.01 + 16.00)

= 29.7455… g

= 30 g (to 2 sig. fig.)

- 4 marks: calculates correct mass **and** correct significant figures **and** correct unit
- 3 marks: calculates correct mass (incorrect significant figures/missing unit)
- 2 marks: calculates mass of ethanoic acid
- 1 mark: calculates MM ethanoic acid **and/or** n(ethanoic acid) = n(ethanamide)

Test 10: Polymers

Multiple-choice solutions

Question 1

B   Condensation polymerisation

There is an ester linkage formed and it is a repeating structure denoted by the 'n' outside the square bracket. Addition polymers do not have an ester linkage (**A**). There is no such process called esterification polymerisation (**C**). Saponification does not produce a polymer (**D**).

Question 2

D   Polyethylene terephthalate (PET)

The section of the polymer shown is PET.

Question 3

C

Tetrafluoroethylene or tetrafluoroethene is used to make polytetrafluoroethylene (PTFE), commonly known as Teflon. **A** and **B** are used to make PVC and polystyrene respectively and **D** can be used to make polydifluoroethylene.

Question 4

B   Chloroethene

A is incorrect because it does not contain chlorine, and **C** and **D** are not IUPAC names.

Question 5

B

The amino acid must have an amino group ($-NH_2$) at one end and a carboxylic acid ($-COOH$) group at the other end.

Question 6

C

C is an addition polymer around the carbon–carbon double bond.

Question 7

D 4500

Because molar mass is $79\,596\,\text{mol}^{-1}$, this is the mass of the polymer composed of a number of monomers. Because the monomer has $(3 \times C + 3 \times H + 1 \times N)$:

$$n(\text{monomer}) = \frac{79\,596}{(3 \times 12.01) + (3 \times 1.008) + 14.01} = 1500\,\text{mol}$$

There are three carbon atoms per monomer; therefore, $3 \times 1500 = 4500$ carbon atoms in the polymer section.

Question 8

C Nylon, PET, HDPE

Nylon has the highest tensile strength, whereas HDPE has the lowest.

Question 9

A Low-density polyethylene (LDPE)

The production of LDPE is through a step-growth polymerisation where termination is a random process. Therefore, a range in molecular masses is found in the polymer produced.

Question 10

A 1 621 418 g

$MM(\text{glucose}) = 180.156$, $MM(\text{water}) = 18.016$

Mass of polymer $= (10\,000 \times 180.156) - (9999 \times 18.016) = 1\,621\,418.016\,\text{g} = 1\,621\,418\,\text{g}$

Short-answer solutions

Question 11

a

Substance	Melting point (°C)
Ethene	−169
Polyethene	110
Polyester	260
Nylon 6,6	269

- 2 marks: correctly matches all substances
- 1 mark: identifies ethene as having the lowest melting point **and** nylon 6,6 as having the highest melting point

b   The melting point of a substance is determined by the strength of intermolecular forces. The weaker the forces, the lower the melting point.

Ethene is a small, non-polar molecule and it has weak dispersion forces between its molecules. Therefore, it will have the lowest melting point.

Nylon 6,6 is a polyamide, which has polar sections within its large chains. In addition to dispersion forces, dipole–dipole and hydrogen bonds can form between the chains of the amide groups.

A polyester has only dipole–dipole forces in addition to dispersion forces. Polyethylene has just dispersion forces, but, being a large molecule, its melting point is higher than that of ethene.

- 2 marks: provides thorough explanation, identifying features of the polymers **and** sketches intermolecular forces
- 1 mark: provides a limited explanation **and/or** sketches intermolecular forces

Question 12

a

$$HO—\underset{\underset{H}{|}}{\overset{\overset{H}{|}}{C}}—\underset{\underset{H}{|}}{\overset{\overset{H}{|}}{C}}—OH \qquad HO—\overset{\overset{O}{\|}}{C}—(CH_2)_2—\overset{\overset{O}{\|}}{C}—OH$$

> • 2 marks: correctly identifies both monomers by drawing structural formulae
> • 1 mark: correctly identifies one monomer by drawing structural formulae **or** correctly identifies both monomers by condensed formulae

b Condensation polymerisation

> • 1 mark: correctly states condensation polymerisation

Question 13

Alkenes have double bonds, which are reactive. In the manufacture of LDPE, an initiator such as benzoyl peroxide is required, whereas in the production of HDPE, a Ziegler–Natta catalyst is required.

> • 2 marks: identifies that alkenes have a reactive double bond **and** states one reagent
> • 1 mark: states that alkenes have a double bond **and/or** one reagent

Question 14

LDPE is composed of branched chains and is therefore more transparent and flexible than HDPE. HDPE has chains that are packed closely together and is opaque and rigid.

> • 2 marks: compares two physical properties **and** relates to structure
> • 1 mark: compares one physical property **and/or** relates to structure

Question 15

$m(PTFE) = $ density $\times$ volume $= 2.2 \times 1.5 \times 10^6 \times 10^3 = 3.3 \times 10^9 \, g$

PTFE is made up of tetrafluoroethylene

$$\underset{F}{\overset{F}{\diagdown}}C = C\underset{\diagdown F}{\overset{\diagup F}{}} \qquad MM = 64.036$$

$n(\text{tetrafluoroethylene}) = \dfrac{m}{MM} = \dfrac{3.3 \times 10^9}{64.036} = 51\,533\,512.4 \, mol$

Because there are four F atoms per molecule, there are $(4 \times 515\,335...) = 206\,134\,049.6 = 2.1 \times 10^8$ fluorine atoms (to 2 sig. fig.).

> • 4 marks: converts volume in L to mL **and** calculates mass of sample using density **and** calculates molar mass of tetrafluoroethylene **and** calculates number of moles of tetrafluoroethylene **and** multiplies number of moles of tetrafluoroethylene by 4 to determine number of fluorine atoms
> • 3 marks: converts volume in L to mL **and/or** calculates mass of sample using density **and** calculates molar mass of tetrafluoroethylene **and** calculates number of moles of tetrafluoroethylene **and** multiplies number of moles of tetrafluoroethylene by 4 to determine number of fluorine atoms
> • 2 marks: calculates number of moles of tetrafluoroethylene correctly
> • 1 mark: calculates *MM* of tetrafluoroethylene **or** mass of sample

Question 16

Polymers that are used to make single-use plastic shopping bags and plastic straws are most likely to be LDPE and these are not biodegradable. These items are used widely by society and contribute to landfill.

Some of these items may even be washed into aquatic environments. Plastic pollution in marine environments threatens aquatic life because many animals eat these items, filling their stomachs with plastic while depriving themselves of food, resulting in their death. These actions have an impact on the food chain.

If these plastic items are destroyed by incineration, they release carbon dioxide, whereas methane gas is released from landfills. Non-stick frying pans are coated with PTFE or Teflon, which is also not biodegradable. But these items are not single use, and so do not contribute to landfill at the same rate as single-use plastic bags and straws.

> An 'identify' question requires you to identify issues and provide points for and/or against.

Question 17

a   Cable ties are relatively cheap compared with the \$2.7 billion rover. The properties of the cable ties enable them to be used in the extreme conditions in space where there are extreme radiation and temperature fluctuations. The cable ties are flexible, so can be used to secure equipment and they do not weather or age rapidly. These factors help in space missions because the cable ties can last for long periods of time.

> • 2 marks: states that cable ties are relatively cheap **and** links their properties to their use
> • 1 mark: states that cable ties are relatively cheap **and/or** links some properties to their use

b $n(\text{ethene}) = n(\text{tetrafluoroethene}) = \dfrac{10 \times 10^6}{24.79} = 4.033\,88\ldots \times 10^5\,\text{mol}$

$n(\text{ETFE}) = 4.033\,88\ldots \times 10^5\,\text{mol}$

The monomers will combine in a $1:1$ ratio.

One polymer unit has $(4 \times \text{C} + 4 \times \text{H} + 4 \times \text{F})$

$$m(\text{ETFE}) = n \times MM$$
$$= 4.033\,88\ldots \times 10^5 \times ((4 \times 12.01) + (4 \times 1.008) + (4 \times 19.00))$$
$$= 5.166\ldots \times 10^7\,\text{g}$$

$m(\text{ETFE}) = 51.66\,\text{tonnes (to 4 sig. fig.)}$

Therefore, the mass of the ETFE will be 51.66 tonnes.

> • 3 marks: calculates correct answer **and** correct significant figures **and** correct unit
> • 2 marks: calculates correct answer **and** correct unit
> • 1 mark: calculates number of moles of ETFE correctly

Test 11: Analysis of inorganic substances

Multiple-choice solutions

Question 1

D Addition of sodium sulfate solution

Lead(II) sulfate would precipitate. **A** is incorrect because lead is toxic and should not be used in a flame test. **B** is incorrect because gravimetric analysis is a quantitative procedure, which is not required for qualitative analysis. **C** will not produce a visible reaction because all nitrates are soluble.

Question 2

C Ligands

The NH_3 groups in the complex ion are ligands. **A** refers to the entire ion. **B** and **D** are not relevant.

Question 3

D Coordinate covalent, covalent

Bond 1 is coordinate covalent in which both electrons are from the nitrogen atom. Bond 2 is covalent in which nitrogen and hydrogen each share one electron.

Question 4

D Addition of nitric acid

Bubbling will be observed when nitric acid is added to an aqueous solution of sodium carbonate but there will be no visible reaction with sodium chloride. **A**, **B** and **C** are incorrect because the metal ion is sodium in both solutions of sodium carbonate and sodium chloride.

Question 5

A The calculated mass of sulfate in the fertiliser is more than the mass of the fertiliser because the precipitate may not have been rinsed with water sufficiently and then not dried to constant mass.

$$n(BaSO_4) = \frac{4.12}{137.3 + 32.07 + 4(16.00)} = 0.017\,67\ldots \text{mol}$$

$$m(SO_4^{2-}) = n \times MM = 0.017\,67\ldots \times (32.07 + 4(16.00)) = 1.698\ldots \text{g}$$

$$m(SO_4^{2-}) > 1.50\,\text{g } m(\text{fertiliser})$$

Therefore, the barium sulfate precipitate may have impurities adhering to it because it wasn't rinsed with water sufficiently and the precipitate may not have been dried to constant mass in the oven.

Question 6

B 2, 1, 3

Spectrum 2 is closest in absorbance to that of pure water so it must be ground water. Spectrum 3 is the least similar to that of pure water and must therefore be contaminated, and must be river water. This leaves spectrum 1 to be tap water.

Question 7

B Green

The solution is green because it is not absorbing in the green region of the spectrum.

Question 8

C 400 nm

Solution 2 absorbs at 400 nm, whereas solution 1 absorbs very little.

Question 9

B 17.8 mg

$$\frac{c(Pb^{2+})}{c(20\ ppm)} = \frac{absorbance\ (0.600)}{absorbance\ (0.300)}$$

$$c(Pb^{2+}) = 40\ ppm$$

In 200 mL, $m(Pb^{2+}) = 40 \times 0.200 = 8\ mg$

$$n(Pb^{2+}) = \frac{8.0 \times 10^{-3}}{207.2} = 3.86... \times 10^{-5}\ mol$$

$$Pb^{2+}(aq) + 2I^-(aq) \rightarrow PbI_2(s)$$

$$n(PbI_2) = n(Pb^{2+}) = 3.86... \times 10^{-5}\ mol$$

$$
\begin{aligned}
m(PbI_2) &= n \times MM \\
&= 3.86... \times 10^{-5} \times (207.2 + (2 \times 126.9)) \\
&= 1.779... \times 10^{-2}\ g = 1.779... \times 10^{-2} \times 10^3 \\
&= 17.79\ mg \\
&= 17.8\ mg\ (to\ 3\ sig.\ fig.)
\end{aligned}
$$

Question 10

C 0.48%

At 0.50, $c(MnO_4^-) = 25\ mg\,L^{-1}$ in diluted sample, i.e. 25 mg in 1000 mL.

In 100 mL diluted sample, there will be 2.5 mg of MnO_4^-.

In 1.00 L of undiluted sample, there will be $50 \times 2.5\ mg = 125\ mg$.

$$n(MnO_4^-) = \frac{m}{MM} = \frac{125 \times 10^{-3}}{54.94 + 4(16.00)} = 0.001\,050\,95\ mol$$

From the equation: $n(Mn^{2+}) = n(MnO_4^-) = 0.001\,050\,95$

$$m(Mn^{2+}) = n \times MM = 0.001\,050\,95 \times 54.94\ g = 0.057\,73\ g$$

$$\%Mn = \frac{0.057\,73...}{12.0} \times 100 = 0.481\%$$

$$\%Mn = 0.48\%\ (to\ 2\ sig.\ fig.)$$

Short-answer solutions

Question 11

Monitoring requires collection of data – both quantitative and qualitative – that results in evidence for making changes as necessary to maintain optimum conditions.

Environmental monitoring serves a vital scientific role by maintaining optimum conditions for flora and fauna.

Heavy metals such as lead ions and mercury ions need to be monitored because of their toxicity and biomagnification/bioaccumulation. AAS can be used.

Trace elements such as copper and zinc are required in small concentrations to ensure healthy growth of plants. Again AAS can be used.

An imbalance in the phosphate-to-nitrate ratio results in eutrophication. Colourimetry can be used to determine phosphate levels by reacting the sample with ammonium molybdate, which produces a bright yellow phosphorus molybdate complex. Maximum complex absorption is at 340 nm. It is proportional to the concentration of inorganic phosphate in the sample.

Calcium and/or magnesium ions can contribute to hard water. AAS can be used.

- 5 marks: identifies that monitoring requires collection of data **and** identifies the need for environmental monitoring **and** identifies at least three ions that need to be monitored **and** identifies environmental issues related to the three ions **and** states analytical techniques used to identify named ions
- 4 marks: identifies that monitoring requires collection of data **and/or** identifies the need for environmental monitoring **and** identifies at least three ions that need to be monitored **and** identifies environmental issues related to the three ions **and** states analytical techniques used to identify named ions
- 3 marks: identifies that monitoring requires collection of data **and/or** identifies the need for environmental monitoring **and** identifies two or three ions that need to be monitored **and** identifies environmental issues related to the two or three ions **and** states analytical techniques used to identify named ions
- 2 marks: identifies that monitoring requires collection of data **and/or** identifies the need for environmental monitoring **and** identifies one or two ions that need to be monitored **and** identifies environmental issues related to the one or two ions **and** states analytical techniques used to identify named ions
- 1 mark: identifies that monitoring requires collection of data **and/or** identifies the need for environmental monitoring **and/or** identifies some ions that need to be monitored **and/or** identifies environmental issues related to the ions **and/or** states analytical techniques used to identify named ions

Question 12

1. Label the bottles X, Y and Z.

2. Label three test tubes X, Y and Z.

3. Pour 1 mL of each solution into the appropriate test tube and test with litmus.

4. If red litmus turns blue, then the solution contains CH_3COO^-.

5. If litmus stays red/blue, then the solution contains I^- or PO_4^{3-}.

6. Add 1 mL lead(II) nitrate solution to the remaining two solutions.

7. If a yellow precipitate forms, the solution contains I^-; if a white precipitate forms, the solution contains PO_4^{3-}.

$$Pb^{2+}(aq) + 2I^-(aq) \rightarrow PbI_2(s)$$

- 5 marks: labels samples **and** states quantities used in tests **and** outlines a logical sequence **and** describes relevant observations **and** states one correct chemical equation with states
- 3–4 marks: labels samples **and/or** states quantities used in tests **and** outlines a logical sequence **and/or** describes relevant observations **and** states one correct chemical equation with states
- 1–2 marks: labels samples **and/or** states quantities used in tests **and/or** outlines a logical sequence **and/or** describes relevant observations **and/or** states one correct chemical equation with states

Question 13

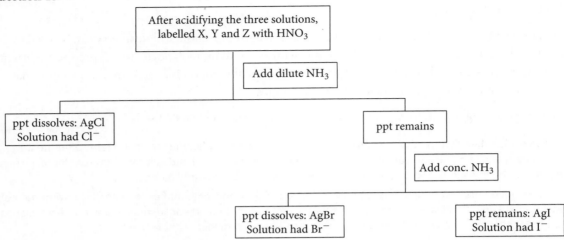

- 3 marks: labels the samples **and** draws a flow chart with a logical sequence to identify all three ions **and** states observations **and** writes one equation to show formation of the complex ion
- 2 marks: draws a flow chart with a logical sequence to identify some of the ions **and/or** states observations **and/or** writes one equation to show formation of the complex ion
- 1 mark: draws a flow chart with a logical sequence to identify some of the ions **or** writes one equation to show formation of the complex ion

Question 14

$n(\text{AgNO}_3)$ added $= cV = 0.101 \times 0.0500 = 0.00505\,\text{mol}$

$n(\text{SCN}^-) = cV = 0.105 \times 0.03872 = 0.0040656\,\text{mol}$

$n(\text{Ag}^+)_{\text{INXS}} = n(\text{SCN}^-) = 0.0040656\,\text{mol}$

$n(\text{Ag}^+)$ reacted $= n(\text{Ag}^+)$ added $- n(\text{Ag}^+)_{\text{INXS}} = 0.00505 - 0.0040656 = 0.0009844\,\text{mol}$

(Cl^-) in $25\,\text{mL}$ sample $= n(\text{Ag}^+)$ reacted $= 0.0009844\,\text{mol}$

$n(\text{Cl}^-)$ in $250\,\text{mL}$ sample $= 10 \times 0.0009844 = 9.844 \times 10^{-3}\,\text{mol}$

$m(\text{Cl}^-) = n \times MM = 9.844 \times 10^{-3} \times 35.45 = 0.3489698\,\text{g} = 0.3489698 \times 1000\,\text{mg} = 348.96\ldots\,\text{mg}$

Because all the Cl^- came from the original $25.00\,\text{mL}$ tomato sauce sample:

mass of the original sample $=$ density $\times$ volume $= 1.20 \times 25.00 = 30.00\,\text{g}$

$c(\text{Cl}^-)$ in ppm $= \dfrac{\text{mg NaCl}}{\text{kg solution}} = \dfrac{348.96\ldots\text{mg}}{0.030\,\text{kg}} = 11632.326\ldots = 1.16 \times 10^4\,\text{ppm}$ (to 3 sig. fig.)

- 5 marks: correct answer **and** correct significant figures **and** correct unit
- 4 marks: correct answer **and** one of correct significant figures **or** correct unit **or** correct answer in grams
- 3 marks: one minor mathematical error **or** calculates mass of chloride in grams correctly
- 1–2 marks: calculates number of moles of chloride ions in sample **and/or** calculates number of moles of silver ions reacted

Question 15

a

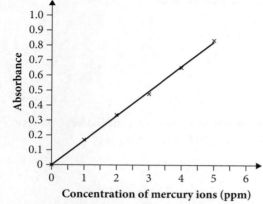

> - 3 marks: draws a line of best fit starting at zero **and** correct *x*- and *y*-axes **and** appropriate unit for *x*-axis **and** appropriate scale
> - 2 marks: draws a line of best fit starting at zero **and** correct *x*- and *y*-axes **and/or** appropriate unit for *x*-axis **and/or** appropriate scale
> - 1 mark: draws a line of best fit **and/or** correct *x*- and *y*-axes **and/or** appropriate unit for *x*-axis **and/or** appropriate scale

b 5.5 ppm

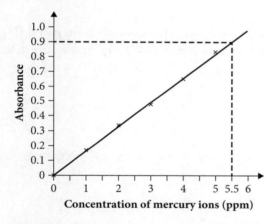

c The concentration of 5.5 ppm is neither reliable nor valid.

It is not reliable because the data points for the standard solutions do not form a line. A line of best fit had to be drawn to account for errors.

The concentration for absorbance of 0.900 had to be determined by extrapolation. According to the Beer–Lambert law, absorbance is not necessarily directly proportional to concentration at high concentrations.

The extrapolation assumed that the same trend continues for high concentrations.

> - 3 marks: states that not reliable and not valid **and** states detailed reasons with reference to the data
> - 2 marks: states that not reliable and not valid **and** states limited reasons with reference to the data
> - 1 mark: states that not reliable and not valid **and/or** states limited reasons with reference to the data

Test 12: Analysis of organic substances

Multiple-choice solutions

Question 1

A Infrared spectrum

An infrared spectrum has wavenumbers along the horizontal axis. A mass spectrum (**B**) has m/z ratio. An NMR spectrum (**C**) has shift in ppm and a UV–vis spectrum (**D**) has wavelengths.

Question 2

C

The trough at about $1750 \, \text{cm}^{-1}$ is due to the carbonyl group, C=O.

The absence of a broad trough at 3230–$3500 \, \text{cm}^{-1}$ shows the absence of a hydroxyl group, OH, so **B** and **D** are incorrect. The absence of a very broad trough at 2500–$3000 \, \text{cm}^{-1}$ shows the absence of a carboxylic acid group, so **A** is incorrect.

Question 3

C No, yes, yes

It will not react with acidified $KMnO_4$ because the hydroxyl group is on a tertiary carbon, so **A** and **D** are incorrect. The hydroxyl group will undergo a substitution reaction with HBr, whereas the carboxylic acid groups will react with sodium carbonate solution.

Question 4

A C_5H_{12}

Because the compound is a hydrocarbon, it contains only carbon and hydrogen, and the molar mass of **A** is 72, which corresponds to the parent ion at 72.

Question 5

B 43

The tallest peak is the base peak at m/z 43.

Question 6

D 2-Bromopropane

The peaks observed at 0 ppm are due to the solvent TMS. There are only two carbon environments because 2-bromopropane is symmetrical, so two peaks are observed in the ^{13}C NMR spectrum. The splitting pattern can be explained because the H labelled 'A' will be a septet while the H labelled 'B' will be a doublet.

$$\begin{array}{c} \text{A} \\ \text{H} \\ | \\ CH_3 - C - CH_3 \\ \text{B} \quad | \quad \text{B} \\ Br \end{array}$$

Question 7

C 1,2,3-Trichloropropane

Possible combinations of masses are given in the table for a compound containing three chlorine atoms, given chlorine can have isotopes of mass 35 and 37.

A and **B** contain one and two chlorine atoms and so would not produce the molecular ion peaks at the large masses stated in the question.

$(3 \times C + 5 \times H) = (3 \times 12 + 5 \times 1) = 41$

C + H	Cl	Cl	Cl	Total
41	35	35	35	146
41	35	35	37	148
41	35	37	37	150
41	37	37	37	152

Question 8

A 1, 2, 3

The W protons have two neighbouring protons so will be split into a triplet. The X protons have three neighbouring protons so will be split into a quartet. The Y proton has no neighbouring protons and will be a singlet.

Question 9

C 6

The six different proton environments are shown in the diagram.

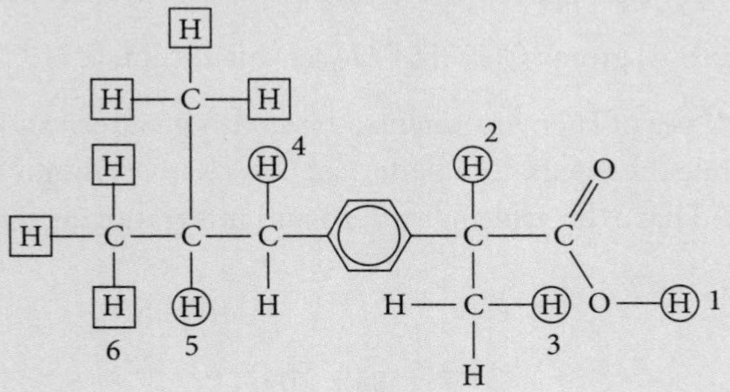

Question 10

A C_5H_8

$m(Br_2) = \text{density} \times \text{volume} = 1.307 \times 15.283 = 19.975\,g$

$n(Br_2) = \dfrac{m}{MM} = \dfrac{19.975}{2 \times 79.90} = 0.125\,mol$

$n(\text{unknown}) = 0.125\,mol$

$m(\text{unknown}) = \text{density} \times \text{volume} = 0.640 \times 13.304 = 8.514\,56\,g$

$MM(\text{unknown}) = \dfrac{m}{n} = \dfrac{8.514\,56}{0.125} = 68.116\,89\,g$

The ^{13}C NMR spectrum shows five different carbon environments. Therefore, the likely formula of the unsaturated hydrocarbon is C_5H_8.

Short-answer solutions

Question 11

Analytical testing can be chemical or spectroscopic.

Analytical tests include the use of bromine water to distinguish a saturated hydrocarbon from an unsaturated hydrocarbon. For example, to distinguish cyclohexane from cyclohexene, bromine water can be used in the absence of UV light. Cyclohexene will decolourise the bromine water because of the presence of the double bond, but the cyclohexane will not.

Equation:

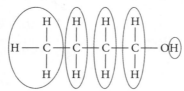

Cyclohexene   1,2-Dibromocyclohexane

Another chemical test is to use sodium carbonate to detect the carboxylic acid group because bubbling will be observed due to evolution of carbon dioxide gas.

Equation:

$$CH_3COOH(l) + Na_2CO_3(aq) \rightarrow NaCH_3COO(aq) + CO_2(g) + H_2O(l)$$

Mass spectroscopy provides information about the molar mass from the parent or molecular ion, while the mass fragments show the make-up of the compound or any halogen isotopes. For example, the mass spectrum of pentane, C_5H_{12}, would show a parent ion at 72 with fragments at 57, 43 and 29 due to fragmentation of the parent ion as shown:

$72 \rightarrow 57$ is due to the loss of a methyl group, CH_3, leaving the ion $[CH_3CH_2CH_2CH_2]^+$, which has a mass of 57.

$72 \rightarrow 43$ is due to the loss of an ethyl group, CH_2CH_3, leaving the ion $[CH_3CH_2CH_2]^+$, which has a mass of 43.

$72 \rightarrow 29$ is due to the loss of a propyl group, $CH_2CH_2CH_3$, leaving the ion $[CH_3CH_2]^+$, which has a mass of 29.

A ^1H NMR spectrum shows types of H environments; the splitting pattern shows arrangement of H atoms on neighbouring carbons, while the integration pattern shows relative amounts of H in each environment. For example, butan-1-ol would have the splitting pattern and integration as shown in the diagram.

Splitting pattern	Triplet	Sextet	Quintet	Triplet	Singlet
Integration	3	2	2	2	1

A ^{13}C NMR spectrum shows types of C environments. For example, butan-1-ol has four carbon environments so four peaks would be seen.

UV–vis spectroscopy shows types of bonds present, e.g. double bonds.

IR radiation is absorbed by bonds to vibrate, bend or stretch. It can help identify the types of functional groups, e.g. OH in alcohols produces a broad trough, whereas OH in carboxylic acids produces a very broad trough. The fingerprint region is below $500–1500 \, cm^{-1}$ and each compound has its unique spectrum in this region.

- 7–8 marks: identifies that analytical testing can be chemical or spectroscopic **and** describes thoroughly three processes **and** identifies three different functional groups **and** states three compounds with these different functional groups **and** describes the effect of these functional groups on the processes **and** states relevant chemical equations where appropriate

- 5–6 marks: identifies that analytical testing can be chemical or spectroscopic **and** describes in a sound manner three processes **and** identifies three different functional groups **and** states three compounds with these different functional groups **and** describes the effect of these functional groups on the processes **and** states relevant chemical equations where appropriate

- 3–4 marks: identifies that analytical testing can be chemical or spectroscopic **and/or** describes in a limited manner three processes **and/or** describes thoroughly two processes **and/or** identifies two or three different functional groups **and/or** states two or three compounds with these different functional groups **and** describes the effect of these functional groups on the processes **and** states relevant chemical equations where appropriate

- 1–2 marks: states names of some analytical tests **and/or** states the relevance of the named processes to identified compounds

Question 12

a TMS, tetramethylsilane, is used as a solvent in NMR spectroscopy. Its structure is shown. TMS is useful because it is symmetrical and has 12 hydrogen atoms and 4 carbon atoms, all of which are in exactly the same environment. They are joined to exactly the same atoms in exactly the same way. This produces a tall single peak.

$$CH_3 - \underset{\underset{CH_3}{|}}{\overset{\overset{CH_3}{|}}{Si}} - CH_3$$

The electrons in the C–H bonds are closer to the hydrogen atoms, which shield the hydrogen nuclei from the external magnetic field. The magnetic field is the strongest to bring the hydrogen atoms back into resonance. This means most other compounds produce peaks to the left of TMS.

- 3 marks: states TMS, tetramethylsilane, **and** draws structure of TMS **and** identifies that TMS is symmetrical with 12 H and 4 C, which are all in exactly the same environment **and** most compounds produce peaks to the left of TMS

- 2 marks: states TMS or tetramethylsilane, **and/or** draws structure of TMS **and** identifies that TMS is symmetrical with 12 H and 4 C, which are all in exactly the same environment **and/or** most compounds produce peaks to the left of TMS

- 1 mark: states TMS or tetramethylsilane, **or** draws structure of TMS **or** identifies that TMS is symmetrical with 12 H and 4 C, which are all in exactly the same environment **or** most compounds produce peaks to the left of TMS

b The horizontal scale is called the chemical shift and is measured in parts per million (ppm).

A peak at a chemical shift of, say, 5.0 means that the hydrogen atoms that caused that peak need a magnetic field five-millionths less than the field needed by TMS to produce resonance. The peak is described as being downfield of TMS. The further to the left, the more downfield the peak is.

The further to the left, i.e. the larger the ppm, the less shielded is the nucleus, e.g. a shift of 10 ppm needs a magnetic field ten-millionths less than TMS. This would happen when the atom is bonded to a more electronegative atom.

- 3 marks: identifies what the horizontal scale shows, including its unit (i.e. ppm) **and** explains chemical shift in terms of shielding **and** links to electronegativity of atom attached to it

- 2 marks: identifies what the horizontal scale shows **and/or** states its unit (i.e. ppm) **and/or** identifies that chemical shift is linked to shielding **and/or** links to electronegativity of atom attached to it

- 1 mark: identifies what the horizontal scale shows **and/or** states its unit, i.e. ppm

Question 13

Infrared spectroscopy would be able to distinguish between the three compounds.

The hydroxyl group in ethanol would show a broad trough at 3230–$3550\,\text{cm}^{-1}$.

The hydroxyl group in ethanoic acid would show a very broad trough at 2500–$3000\,\text{cm}^{-1}$.

There is no hydroxyl group in ethyl ethanoate, so only a peak at 1680–$1750\,\text{cm}^{-1}$ due to the carbonyl $C{=}O$ group in the ester would be observed.

The student's statement is therefore valid because infrared spectroscopy would help to distinguish between the compounds listed.

Question 14

The empirical formula is calculated as shown.

	C	**H**	**N**
%	61	15	24
Mass (g) in 100 g	61	15	24
$n = \dfrac{m}{MM}$	$\dfrac{61}{12.01}$	$\dfrac{15}{1.008}$	$\dfrac{24}{14.01}$
n	5.08	14.9	1.71
Divide by smallest no.	3.0	8.7	1
	3	9	1

Empirical formula is C_3H_9N

From the mass spectrum, the parent ion has a mass of 59.

$59.112 \times n = 59$

$\qquad n = 1$

Hence, molecular formula is C_3H_9N.

Must be an amine because it reacts with HCl to produce an ammonium salt.

From the IR spectrum, the compound is an amine, e.g. absence of $C{=}O$ trough, presence of N–H trough at 3300–$3500\,\text{cm}^{-1}$.

From the ^{1}H NMR spectrum, there are 3 H environments, as shown.

From the ^{13}C NMR spectrum, there are 2 C environments, as shown.

The compound is propan-2-amine.

- 7 marks: states correct IUPAC name **and** correct structural formula **and** calculates correct empirical formula **and** refers to relevant spectra for data, including mass spectrum and NMR spectrum
- 5–6 marks: states correct IUPAC name **and/or** correct structural formula **and** calculates correct empirical formula **and** refers to relevant spectra for data, including mass spectrum and NMR spectrum
- 3–4 marks: states correct IUPAC name **and/or** correct structural formula **and** calculates correct empirical formula **and** refers to relevant spectra for data
- 1–2 marks: states correct IUPAC name **and/or** correct structural formula **and/or** calculates correct empirical formula **and/or** refers to relevant spectra for data

Test 13: Chemical synthesis and design

Multiple-choice solutions

Question 1

B It is amphoteric.

Aluminium oxide is amphoteric because it reacts with basic sodium hydroxide. However, it does not contain a proton so it is not amphiprotic.

Question 2

C Carbon is oxidised and donates electrons. Aluminium ions are reduced and gain electrons.

From the overall equation, it can be seen that the oxidation number of aluminium is reduced from +3 to 0 while the oxidation number of carbon increases from 0 to +4. Aluminium ions are therefore reduced at the cathode while carbon is oxidised at the anode.

Question 3

D 63.0%

$$n(Al) = \frac{m}{MM} = \frac{750\,000}{26.98} = 2.7798\ldots \times 10^4\,mol$$

$$n(Al_2O_3) = \frac{n(Al)}{2} = 1.3899\ldots \times 10^4\,mol$$

$$m(Al_2O_3) = n \times MM = 1.3899\ldots \times 10^4 \times ((2 \times 26.98) + (3 \times 16.00)) = 1.417\ldots \times 10^6\,g$$

$$\%\ purity = \frac{m(aluminium\ oxide)}{m(bauxite)} \times 100 = \frac{1.417 \times 10^6}{2.25 \times 10^6} \times 100 = 62.98\ldots = 63.0\%\ (to\ 3\ sig.\ fig.)$$

Question 4

C Location 3 is the preferred location because it is close to the town where employees may live, and has road and railway access, and has access to hydroelectric power for the electrolysis process.

Question 5

A 293 g

$1000 \times 0.99 = 990\,g$

$990 \times 0.37 = 366.3\,g$

$366.3 \times 0.8 = 293.04\,g$

Question 6

C 81%

$$n(\text{methanol}) = \frac{m}{MM} = \frac{515.0}{12.01 + (4 \times 1.008) + 16.00} = 16.07\ldots \text{mol}$$

$$n(\text{ethanoic acid}) = \frac{m}{MM} = \frac{515.0}{(2 \times 12.01) + (4 \times 1.008) + (2 \times 16.00)} = 8.575\ldots \text{mol}$$

Limiting reagent is ethanoic acid.

$n(\text{methyl ethanoate}) = n(\text{ethanoic acid}) = 8.575\ldots \text{mol}$

$m(\text{methyl ethanoate})$ for 100% yield $= n \times MM = 8.575\ldots \times MM$

$$= 8.575\ldots \times (3 \times 12.01) + (6 \times 1.008) + (2 \times 16.00)$$

$$= 635.285\,5858 \text{ g}$$

$$\% \text{ yield} = \frac{515.0}{635.28\ldots} \times 100 = 81.07\%$$

Question 7

B Reduction of the amount of ammonium nitrate stored

Reducing the amount of nitrate stored minimises the risk of explosions.

Question 8

C 53.80%

$$\% \text{ atom ecomony} = \frac{MM(CH_2Cl_2)}{MM(CH_2Cl_2) + (2 \times MM(HCl))} \times 100$$

$$= \frac{12.01 + (2 \times 1.008) + (2 \times 35.45)}{12.01 + (2 \times 1.008) + (2 \times 35.45) + (2 \times (1.008 + 35.45))}$$

$$= 53.80\%$$

$$CH_4(g) + 2Cl_2(g) \rightarrow CH_2Cl_2(g) + 2HCl(g)$$

Question 9

C Bubble the HCl(g) through water to produce useful hydrochloric acid.

This ensures both products are used.

Question 10

B Availability of sunlight

Availability of sunlight is not usually considered when planning the location of a chemical factory, whereas all other options are considered.

Short-answer solutions

Question 11 ©NESA | 2011 SAMPLE ANSWERS SII Q30

Step 1 in the process is very exothermic as increasing the temperature will force the reaction in the reverse direction. However, as indicated, the reaction occurs at 900°C. This high temperature is necessary to keep up the rate of reaction.

To compensate for the high temperature, it would be appropriate to decrease the pressure. This would force the reaction in the forward direction, as there are more moles of product than moles of reactants.

Removing NO as it forms would speed up the reaction. Catalysts should be used in all steps to also speed up the reaction.

In step 2, the temperature isn't as critical because the ΔH is only -114 kJ. NO_2 is more stable than NO in the presence of O_2. For this reaction, an increase in pressure is necessary in order to force the reaction to the product side, producing more NO_2. NO_2 should also be removed as it forms to keep the reaction proceeding in the forward direction.

Step 3 is not an equilibrium reaction; therefore, increasing the temperature will increase the rate of reaction, producing more HNO_3 despite it also being an exothermic reaction.

- 6 marks: demonstrates a thorough knowledge and understanding of equilibrium and Le Chatelier's principle with reference to the THREE reactions. Relates the conditions required to the increased yield and production rate. Demonstrates coherence and logical progression and includes correct use of scientific principles and ideas
- 4–5 marks: demonstrates a sound knowledge and understanding of equilibrium and Le Chatelier's principle with reference to the three reactions. Communicates some scientific principles and ideas in a clear manner
- 2–3 marks: demonstrates a basic knowledge and understanding of equilibrium and Le Chatelier's principle with reference to flow chart. Communicates ideas in a basic form using general scientific language
- 1 mark: demonstrates a limited knowledge and understanding of equilibrium and Le Chatelier's principle. Communicates simple ideas

Question 12

$MM(Fe_9Ni_9S_8) = (9 \times 55.85) + (9 \times 58.69) + (8 \times 32.07) = 1287.42\,\text{g mol}^{-1}$

$T = 290 + 273.15 = 563.15\,\text{K}$

$R = 8.314\,\text{J mol}^{-1}\text{K}^{-1}$

$n(Fe_9Ni_9S_8) = \dfrac{m}{MM} = \dfrac{1.88 \times 10^6}{1287.42} = 1.7166\ldots \times 10^3\,\text{mol}$

For 100% yield, $n(SO_2) = n(Fe_9Ni_9S_8) \times 8 = 1.4602\ldots \times 10^3\,\text{mol}$

Since yield is 70.0%, $n(SO_2) = 1.4602\ldots \times 10^3 \times 0.700 = 1.022\ldots \times 10^3\,\text{mol}$

$V(SO_2) = \dfrac{nRT}{P} = \dfrac{1.022\ldots \times 10^3 \times 8.314 \times 563.15}{160.0} = 2.991\ldots \times 10^5\,\text{L} = 2.99 \times 10^5\,\text{L}$ (to 3 sig. fig.)

- 6 marks: provides correct answer **and** correct significant figures **and** correct unit
- 5 marks: provides correct answer **and** one of correct significant figures **or** correct unit
- 4 marks: calculates number of moles of SO_2 for 100% yield
- 3 marks: calculates number of moles of SO_2 for 70% yield
- 1–2 marks: calculates MM of $Fe_9Ni_9S_8$ **and/or** number of moles of $Fe_9Ni_9S_8$

Question 13

a 350°C and 400 atm are the optimum conditions for maximum yield of about 68%.

> - 2 marks: states correct temperature with unit **and** states correct pressure with unit
> - 1 mark: states correct temperature **and/or** states correct pressure

b $n(N_2) = \dfrac{m}{MM} = \dfrac{1.50 \times 10^3}{2 \times 14.01} = 53.533\,19\,\text{mol}$

$n(H_2) = \dfrac{m}{MM} = \dfrac{700.0}{2 \times 1.008} = 347.2\ldots\,\text{mol}$

Limiting reagent is N_2

For 100% yield:

$n(NH_3) = 2 \times n(N_2) = 2 \times 53.533\,19 = 1.0706\ldots \times 10^2\,\text{mol}$

$m(NH_3) = n \times MM = 1.0706\ldots \times 10^2 \times (14.01 + (3 \times 1.008)) = 1.823\ldots \times 10^3\,\text{g}$

At 450°C and 200 atm, the yield is about 26% as shown in the graph.

$m(NH_3) = 1.823\ldots \times 10^3 \times 0.26 = 4.7418\ldots \times 10^2 = 474\,\text{g}$ (to 3 sig. fig.)

> - 4 marks: calculates correct mass of ammonia produced **and** correct significant figures **and** correct unit
> - 3 marks: calculates correct mass of ammonia produced **and/or** correct significant figures **or** correct unit
> - 2 marks: calculates the limiting reagent to be nitrogen **and/or** calculates mass of ammonia for 100% yield
> - 1 mark: calculates number of moles of nitrogen **and** calculates number of moles of hydrogen

c **i** $H_2O(l) \rightarrow 2H_2(g) + O_2(g)$

ii Site 3 is the most suitable site to produce green ammonia because electricity for the electrolysis of water can be obtained from hydroelectric power. The site is within 40 km of town with road and rail connection for workers to travel to the plant.

Site 1 is not as suitable because the power source for electricity is further away. It is not desirable to use coal power because that produces CO_2.

Site 2 is also not desirable for the same reasons as for site 1.

Site 4 is also not suitable. Although electricity production using nuclear power does not generate CO_2, the nuclear waste needs to be stored safely.

> - 3 marks: identifies that site 3 is the most suitable site **and** states valid reasons **and** states why sites 1, 2 and 4 are not suitable
> - 2 marks: identifies that site 3 is the most suitable site **and/or** states valid reasons **and/or** states why sites 1, 2 and 4 are not suitable
> - 1 mark: identifies that site 3 is the most suitable site **and** states one valid reason for its suitability

Practice HSC exam 1

Multiple-choice solutions

Question 1

A N,N-Dimethylmethanamine

The longest chain has one carbon with two methyl groups attached to the N.

Question 2

B $+150 \, \text{kJ mol}^{-1}$

$\Delta H = H_{\text{products}} - H_{\text{reactants}} = 250 - 100 = +150 \, \text{kJ mol}^{-1}$

Question 3

D Ethanol + ethanoic acid + concentrated sulfuric acid

An ester is made in this reaction and is immiscible with the reaction mixture.

A would produce ethanoic acid, which would be miscible with ethanol. **B** would produce sodium ethoxide, which would also be miscible. **C** would also be miscible and would not react.

Question 4

B Hydrobromic acid, hypobromous acid, bromous acid, bromic acid

A binary acid (HBr) is named with the prefix 'hydro', first syllable of the anion and suffix '-ic'.

A polyatomic anion (bromate, BrO_3^-) is named with the first syllable of the anion and suffix '-ic'. When there is one fewer oxygen in the polyatomic ion (BrO_2^-) it is given the suffix '-ous'. When there are two fewer oxygens (BrO^-) the prefix '-hypo' and suffix '-ous' is given.

Question 5

C 4

The isomers are butan-1-ol, butan-2-ol, 2-methylpropan-1-ol and 2-methylpropan-2-ol.

Question 6

C 7.27

$2.93 \times 10^{-15} = [H^+][OH^-]$

Let $x = [OH^-]$

Since $[H^+] = [OH^-]$

$x^2 = 2.93 \times 10^{-15}$

$x = 5.412\ldots \times 10^{-8}$

$pOH = -\log_{10}[5.412\ldots \times 10^{-8}] = 7.266\,56\ldots = 7.27$

Question 7

D Add dilute ammonia solution. If the precipitate dissolves, it is Cl^-. If the precipitate does not dissolve, add concentrated ammonia. If the precipitate now dissolves, it is Br^-.

When NH_3 is added to insoluble $AgCl$, the ammonia binds to the Ag^+ ion, forming the complex diamine silver, $[Ag(NH_3)_2]^+$, and producing a colourless solution:

$$AgCl(s) + 2NH_3(aq) \rightleftharpoons [Ag(NH_3)_2]^+(aq) + Cl^-(aq)$$

Silver bromide only dissolves in a concentrated ammonia solution to form the same complex ion of silver:

$$AgBr(s) + 2NH_3(aq) \rightleftharpoons [Ag(NH_3)_2]^+(aq) + Br^-(aq)$$

Silver iodide does not dissolve in either dilute or concentrated ammonia solution.

Question 8

C It is not valid because the sample should have been made basic first to ensure only barium phosphate precipitated.

Barium phosphate precipitates under basic conditions only, whereas barium sulfate precipitates under acidic conditions. The procedure is therefore not valid.

Question 9

B Orange colour turns green. 2,2-Dimethylpropanoic acid

The orange-coloured dichromate oxidises the primary alcohol to a carboxylic acid, and green-coloured Cr^{3+} ions form. The name is incorrect in **C** and potassium dichromate is not purple in colour.

Question 10

A 4

Carbons 1 and 4 are not the same because of the difference in their chemical environments. Carbon 1 is $CH_3-C=O$, whereas carbon 4 is CH_3-CH_2. You need to look at the surroundings of the carbon (its environment).

Question 11

B $NH_3(aq) + HCl(g) \rightarrow NH_4Cl(s)$
    base   acid   acid

Ammonia (NH_3) is a base that will accept the donated hydrogen ion from hydrochloric acid (HCl). Ammonium chloride will donate the hydrogen to the chloride ion, acting as a Brønsted–Lowry acid. According to Arrhenius, acid + base produces salt and water; however, in this example, there are no oxygen atoms present in order to produce the water molecule. HCl is a Brønsted–Lowry acid.

Question 12

D Decrease, endothermic

Photosynthesis is an endothermic reaction because more energy is required to form the products than is provided by the reactants. As the reactants form products, the system moves into a more ordered state; therefore, entropy decreases.

Question 13

B $\dfrac{[\text{Fe(SCN)}^{2+}]}{[\text{Fe}^{3+}][\text{SCN}^-]}$

PORK – products over reactants = K

The equation of iron(III) thiocyanate at equilibrium is $\text{Fe}^{3+}(aq) + \text{SCN}^-(aq) \rightleftharpoons \text{Fe(SCN)}^{2+}(aq)$

Question 14

C To sort the ions by mass using acceleration and deflection

The electron beam produces ions by bombarding the vapour – if an electron is added, a negative ion is formed, and if an electron is removed, a positive ion is formed. Neutral particles are not detected in mass spectrometry. The ions travel through a magnetic field and are accelerated or deflected, with small positive ions moving faster than larger ions. The ions reach the detector one after another, based on mass, with lighter ions arriving before heavier ions.

Question 15

B

Hydrobromic acid is a strong acid, whereas calcium hydroxide is a strong base.

Graph **A** represents a strong acid with a weak base. Graph **C** represents a weak acid with a strong base. Graph **D** cannot occur.

Question 16

D No, $Q < K_{sp}$

$[\text{Ba}^{2+}]$:

$$c_1 V_1 = c_2 V_2$$
$$0.05 \times 0.1 = c_2 \times 0.2$$
$$c_2 = 0.025 \, \text{mol L}^{-1}$$

$[\text{OH}^-]$:

$$c_1 V_1 = c_2 V_2$$
$$0.1 \times 0.1 = c_2 \times 0.2$$
$$c_2 = 0.05 \, \text{mol L}^{-1}$$

$$Q = [\text{Ba}^{2+}][\text{OH}^-]^2$$
$$= [0.025][0.05]^2$$
$$= 6.25 \times 10^{-5}$$

Since $Q < K_{sp}$, no precipitate will form.

Question 17

A Ethyl acetate

Although it may be time consuming, drawing the structures can help you visualise the answer.

Question 18

C 3.15

A and B have not taken into account K_a of CH_3COOH.

$$CH_3COOH(aq) + NaOH(aq) \rightarrow NaCH_3COO(aq) + H_2O(l)$$

Mole ratio: x : x : x : x

$n(CH_3COOH) = cV = 0.100 \times 0.100 = 0.010 \, mol$

$n(NaOH) = cV = 0.10 \times 0.050 = 0.005 \, mol$

CH_3COOH is in excess. $n(CH_3COOH)_{INXS} = 0.010 - 0.005 = 0.005 \, mol$

$c(CH_3COOH)_{INXS} = \dfrac{0.005}{0.100 + 0.050} = 0.0333... \, mol$

$$CH_3COOH(aq) \rightarrow H^+(aq) + CH_3COO^-$$

Let $x = [H^+] = [CH_3COO^-]$

$K_a = \dfrac{x^2}{[acid]}$

$1.5 \times 10^{-5} = \dfrac{x^2}{0.0333...}$

$x^2 = 5.0000 \times 10^{-7}$

$x = 7.071\,06... \times 10^{-4} \, mol$

$pH = -\log_{10}(7.071\,06... \times 10^{-4})$

$pH = 3.1512... = 3.15$ (to 2 d.p. for pH)

Question 19

D 2-Methylpropan-2-ol, 2-bromo-2-methylpropane, 2-methylpropene

2-Methylpropan-2-ol has the structure shown.

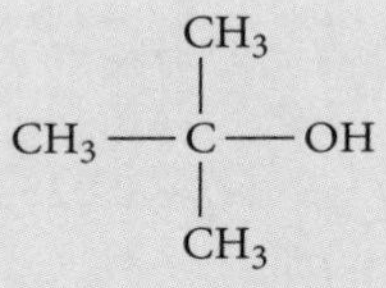

It has two different carbon and two different hydrogen environments as shown. The ratio of the H is $9:1$ in keeping with the ^1H NMR spectrum. It is a tertiary alcohol because it cannot be oxidised with acidified potassium dichromate or potassium permanganate. It undergoes a substitution reaction with HBr and dehydration with concentrated sulfuric acid with subsequent addition reaction with HBr.

A and C would have three signals in the ^{13}C NMR spectra. B would have two signals in the ^{13}C NMR spectrum but would have three signals in the ^1H NMR spectrum.

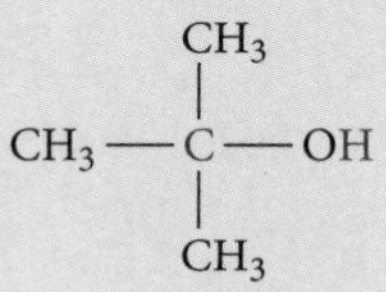

Question 20

D 38%

Absorbance of 0.015 has $c(Cu^{2+})$ of $60 \, mg \, L^{-1}$

Then in 500 mL there is 30 mg Cu^{2+}, which came from 10 mL of the original 1 L. Therefore, in the original 1 L, there is $30 \times 100 \, mg = 3000 \, mg \, Cu^{2+}$

$n(Cu^{2+}) = \dfrac{3000 \times 10^{-3}}{63.55} = 0.0472... \, mol$

$m(Cu^{2+}) = n \times MM = 0.0472... \times 63.55 = 3.000... \, g$

$\% \, Cu^{2+} = \dfrac{m(Cu^{2+})}{m(sample)} \times 100 = \dfrac{3.000...}{5.0} \times 100 = 60\%$

Short-answer solutions

Question 21

a $H_2O(l) \rightleftharpoons H^+(aq) + OH^-(aq)$

b The ionisation of water is an endothermic process.

As temperature increases from 0 to 100°C, pH decreases from 7.47 to 6.14. Therefore, the hydrogen ion concentration increases, and equilibrium shifts to the right, the product side. According to Le Chatelier's principle, when a system is at equilibrium and a change is imposed, the equilibrium shifts to counteract the change, which in this case is in an endothermic direction to use the heat.

- 3 marks: identifies that the ionisation of water is endothermic **and** states Le Chatelier's principle **and** provides explanation by referring to the data provided
- 2 marks: identifies that the ionisation of water is endothermic **and/or** states Le Chatelier's principle **and/or** provides explanation by referring to the data provided
- 1 mark: states one correct point

Question 22

a

H—C—C—C—C—C—C—H with H, H, Br, H, H, H on top and H, H, H, H, H, H on bottom

(Condensed or skeletal drawings are also accepted.)

b H_2O and H_2SO_4 (Any inorganic acid other than HCl is also accepted.)

c

H—C—C with H, H below and O (double bond), OH

(Condensed or skeletal drawings are also accepted.)

Hexan-3-ol can also produce compound D in the presence of acidified permanganate.

d

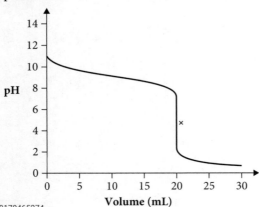

(Expanded or condensed drawings are also accepted.)

Question 23

a weak base/strong acid

b pH 4.5–5

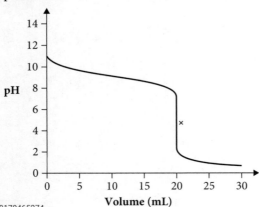

c   When a weak base ionises in water, a strong conjugate acid is produced. As the base is neutralised, this conjugate acid remains ionised in water, therefore decreasing the pH of the neutralised solution.

$$NH_3(aq) + HCl(aq) \rightleftharpoons NH_4Cl(aq)$$
weak base   strong conjugate acid

- 3 marks: links the weak base to the production of a strong conjugate acid ionising in water, therefore decreasing the pH, **and** includes a relevant chemical equation
- 2 marks: links the weak base to the production of a strong conjugate acid ionising in water, therefore increasing the pH (no chemical equation), **or** links the base to the creation of an acidic salt and includes a relevant chemical equation
- 1 mark: links the base to the production of an acidic salt **or** provides a relevant equation **or** provides some relevant information

Question 24

a   The colour will change from yellow-red/orange to a paler yellow as the equilibrium shifts to the left, the reactant side. This is due to the hydroxide ions from NaOH(aq) reacting with the Fe^{3+} ions, causing the concentration of Fe^{3+} ions to decrease as $Fe(OH)_3(s)$ precipitates.

$$Fe^{3+}(aq) + 3OH^-(aq) \rightarrow Fe(OH)_3(s)$$

According to Le Chatelier's principle, if a system is at equilibrium and a change is made that upsets the equilibrium, the system alters to counteract the change and a new equilibrium is established. In this case, equilibrium shifts to the left, the reactant side, to increase the concentration of Fe^{3+} ions.

- 2 marks: thoroughly describes the observation **and** explains the cause of the shift in equilibrium with reference to Le Chatelier's principle **and** writes an equation for precipitation of $Fe(OH)_3$
- 1 mark: describes the observation in a limited manner **and/or** provides a statement about the change **and/or** states Le Chatelier's principle

b

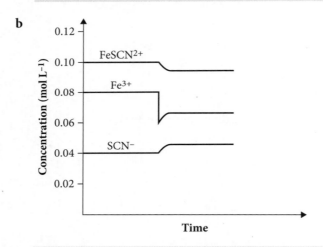

- 3 marks: draws correct change **and** amount of change in a 1 : 1 : 1 ratio **and** correct shape of curve for all three species
- 2 marks: draws correct change **and/or** amount of change in a 1 : 1 : 1 ratio **and** correct shape of curve for one or two species
- 1 mark: provides one correct feature

Question 25

Spectrum 1 is for the alkanoic acid as shown by the very broad O–H peak at 2500–3000 cm^{-1} and the C=O peak at 1700 cm^{-1}.

Spectrum 2 has a C–H peak at 2900–3000 cm^{-1} and a broad O–H peak after 3000 cm^{-1}; therefore, spectrum 2 is for the alcohol.

Spectrum 3 also has peak at 2900–3000 cm^{-1}, indicating a C–H bond, but as there are no other peaks present, spectrum 3 is for the alkane.

- 3 marks: correctly matches all organic compounds to the spectra, using specific reference to the spectrum to justify choice
- 2 marks: correctly matches organic compounds to the spectra
- 1 mark: provides some correct information

Question 26

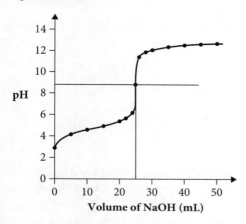

(33 mm = 10 mL, 16 mm = 4.84 mL, V(NaOH) equiv. point = 24.84 mL = 24.8 mL)

$$H_3PO_4(aq) + 3NaOH(aq) \rightarrow Na_3PO_4(aq) + 3H_2O(l)$$

$$n(NaOH) = cV = 0.201 \times 0.0248 = 0.004\,9848 \text{ mol}$$

$$n(H_3PO_4) = \frac{n(NaOH)}{3} = \frac{0.004\,9848}{3} = 0.001\,6616 \text{ mol}$$

$$c(H_3PO_4) = \frac{n}{V} = \frac{0.001\,6616}{0.025\,00} = 0.066464 = 0.0665 \text{ mol L}^{-1} \text{ (to 3 sig. fig.)}$$

- 6–7 marks: provides correct graph with labelled axes **and** identifies equivalence point and volume of NaOH **and** correct balanced equation **and** calculates n(NaOH) **and** calculates n(H$_3$PO$_4$) **and** calculates c(H$_3$PO$_4$) with correct significant figures and unit
- 4–5 marks: as above but incorrect significant figures **or** missing unit **or** correct graph with labelled axes **and** identifies equivalence point and volume of NaOH **and** correct balanced equation **and** calculates n(NaOH) **and/or** calculates n(H$_3$PO$_4$) **and** calculates c(H$_3$PO$_4$)
- 2–3 marks: provides correct graph with labelled axes
- 1 mark: provides some relevant information

Question 27

1. Label the bottles 1–6.

2. Pour 2–3 drops into depression tiles/micro test tubes labelled 1–6.

3. Test with litmus/indicator.

4. If litmus turns blue, then ethanamine.

5. If litmus turns red, then ethanoic acid.

6. Rinse depression tile into a waste bottle for heavy metal collection and dry depression tile.

7. Place 2–3 drops of remaining samples in appropriately labelled depression tiles/micro test tubes.

8. Add 2–3 drops of KI(aq). The test tube in which a bright yellow ppt forms contains lead(II) nitrate.

9. Clean out the precipitate with a paper towel for separate disposal in heavy metal container.

10. Place 1 mL of fresh sample in test tube.

11. Add 1 mL of aqueous sodium hydroxide, heat the mixture gently and test gas with litmus. If it turns blue, it is ethanamide.

12. Place 1 mL of fresh sample in a test tube.

13. Add acidified potassium dichromate solution to the remaining test tube.

14. The test tube in which the colour changes from orange to green contains ethanol.

15. The remaining test tube contains sodium ethanoate.

- 4–5 marks: labels bottles **and** quantities stated **and** procedure is logical with numbered steps **and** uses micro test tubes or depression tiles **and** procedure uses minimal chemicals **and** correct procedure for identifying all substances
- 2–3 marks: labels bottles **and/or** quantities, stated procedure is logical with numbered steps, uses micro test tubes or depression tiles, procedure uses minimal chemicals, correct procedure for identifying some substances
- 1 mark: provides some relevant information

Question 28

Fe^{3+} is added, so the reaction with SCN^- produces blood-red $Fe(SCN)^{2+}$

$$AgNO_3(aq) + KSCN(aq) \rightarrow AgSCN(aq) + KNO_3(aq)$$

$n(AgNO_3)_{added} = cV = 0.1042 \times 0.100 = 0.010\,42 \, mol$

$n(AgNO_3)_{INXS} = n(SCN^-) = 0.053\,51 \times 0.041\,25 = 0.002\,207\,2875 \, mol$

$n(AgNO_3)_{reacted} = n(AgNO_3)_{added} - n(AgNO_3)_{INXS} = 0.010\,42 - 0.002\,207\,2875 = 0.008\,212\,7125 \, mol$

$$Ag^+(aq) + Cl^-(aq) \rightarrow AgCl(s)$$

$n(AgNO_3)_{reacted} = n(Cl^-) = 0.008\,212\,7125 \, mol$

$c(Cl^-) = \dfrac{n}{V} = \dfrac{0.008\,212\,7125}{0.050\,00} = 0.164... \, mol\,L^{-1}$

$c(Cl^-) = 0.164... \times 35.45 \, g\,L^{-1} = 5.8228... \, g\,L^{-1} = 5.8228... \times 1000 \, mg\,L^{-1} = 5823 \, mg\,L^{-1}$ (to 4 sig. fig.)

- 4–5 marks: calculates chloride concentration to correct significant figures and unit **and** identifies species added as indicator
- 2–3 marks: calculation contains one error **and/or** identifies species added as indicator
- 1 mark: provides one correct step

Question 29

$$\Delta H = \frac{q}{n}$$

n(ethanol) $\dfrac{q}{\Delta H} = \dfrac{155 \times 4.18 \times 72}{1\,367\,000} = 0.034\,1249\dots\,\text{mol}$

MM(ethanol) $= (2 \times 12.01) + (6 \times 1.008) + 16.00 = 46.068\,\text{g mol}^{-1}$

m(ethanol) $= n \times MM = 0.034\,1249\dots \times 46.068 = 1.572\dots\,\text{g}$

But if 80% heat is lost, need to burn $1.572\dots\,\text{g}/0.2 = 7.86\,\text{g} = 7.9\,\text{g}$ (to 2 sig. fig.)

- 3 marks: provides correct answer **and** correct significant figures **and** correct unit
- 2 marks: provides correct answer **and/or** one of correct significant figures and correct unit **and/or** calculates mass without heat loss
- 1 mark: calculates number of moles of ethanol correctly

Question 30

a Phosphoric acid is a weak acid and calcium hydroxide is a strong base. A weak acid–strong base titration will result in a basic salt solution. Therefore, the end point of this titration would occur above pH 7. The pH range of bromocresol is pH 4–5. This indicator would be unsuitable because the student would stop the titration too soon, resulting in a calculation of a higher concentration of calcium hydroxide than the actual concentration.

- 4 marks: provides a judgement on the suitability of the indicator, making reference to the pH of the salt solution, range of the indicator and impact on the calculated concentration
- 3 marks: provides a judgement on the suitability of the indicator, making reference to the pH of the salt solution **and/or** range of the indicator **and/or** impact on the calculated concentration
- 2 marks: provides a judgement on the suitability of the indicator, making reference to the range of the indicator **or** impact on calculated concentration **or** provides a judgement on the suitability of the indicator with incorrect strength of acid/base, making reference to the pH of the incorrect salt **and/or** range of the indicator **and/or** impact on the calculated concentration
- 1 mark: provides some relevant information

b Titres A and E should be removed from the average calculation. A is a rough titre, and E is more than ±0.1 from the main result of 11.4 mL.

The results are reliable because three concordant results can be used to calculate the average titre.

- 2 marks: makes a judgement on the reliability of each titrant
- 1 mark: makes a judgement on the collective reliability of the results

c $2H_3PO_4(aq) + 3Ca(OH)_2(aq) \rightleftharpoons Ca_3(PO_4)_2(aq) + 6H_2O(l)$

$n(H_3PO_4) = c \times V = 0.77 \times 0.025 = 0.019\,25\,\text{mol}$

Molar ratio $= 2:3$

$n(Ca(OH)_2) = \dfrac{3}{2} \times 0.019\,25 = 0.028\,875\,\text{mol}$

$c(Ca(OH)_2) = \dfrac{n}{V} = \dfrac{0.028\,875}{0.0114} = 2.5\,\text{mol L}^{-1}$

- 3 marks: correctly determines the concentration of calcium hydroxide to 2 significant figures
- 2 marks: correctly calculates the number of moles of calcium hydroxide
- 1 mark: provides steps towards the calculation of number of moles of calcium hydroxide

Question 31

a $N_2(g) + 3H_2(g) \rightleftharpoons 2NH_3(g)$

	$[N_2]$	$[H_2]$	$[NH_3]$
Initial	4.00	4.00	0
Change	$-x$	$-3x$	$+2x$
	$\dfrac{0.66}{2} = 0.33$	$0.33 \times 3 = 0.99$	$= 0.66$
Equilibrium	$4.00 - 0.33 = 3.67$	$4.00 - 0.99 = 3.01$	0.66
Equilibrium concentration (mol L^{-1})	$\dfrac{3.67}{3} = 1.22$	$\dfrac{3.01}{3} = 1.003$	$\dfrac{0.66}{3} = 0.22$

3 L, so divide all concentrations by 3: $[N_2] = 1.22 \, \text{mol L}^{-1}$, $[H_2] = 1.003 \, \text{mol L}^{-1}$, $[NH_3] = 0.22 \, \text{mol L}^{-1}$

$$K_{eq} = \frac{[NH_3]^2}{[N_2][H_2]^3} = \frac{0.22^2}{1.22 \times 1.003^2} = 0.039$$

- 3 marks: correctly calculates the values of the equilibrium constant
- 2 marks: correctly calculates the concentrations of N_2, H_2 and NH_3 at equilibrium **and** writes correct equilibrium expression
- 1 mark: writes correct equilibrium expression

b   **Increase pressure** – According to Le Chatelier's principle, a system stressed by an increase in pressure will shift to the side with fewer moles, in this case, increasing the yield of ammonia. The higher the pressure, the more costly it is for the manufacturer because it is expensive to purchase and maintain the equipment. Higher pressures also may create safety issues for workers, which will need to be mitigated. Therefore, a manufacturer may decide to double the pressure to increase yield while keeping costs low and safety high.

Removing product – According to Le Chatelier's principle, removing ammonia from the reaction vessel would drive the reaction forward; however, because ammonia is a gas, distillation may be costly and safety concerns arise from gas being released at high temperatures and pressure. Removing the product would also lower pressure, adding to the costs of maintaining a high pressure within the system.

Temperature and a catalyst – The production of ammonia is an exothermic reaction; therefore, according to Le Chatelier's principle, the system will favour a lower temperature in order to drive forward. However, a low temperature would decrease the rate of reaction; so it would take a very long time for the manufacturer to obtain their product, hence increasing production costs. As the iron catalyst is present to increase the rate of reaction, using a higher temperature could increase the rate of reaction enough to increase yield. Increasing the temperature too much would be expensive to maintain and increase safety concerns.

- 6 marks: thoroughly explains, using Le Chatelier's principle, how the reaction conditions of temperature, pressure, concentration and catalyst can be used to increase the yield of ammonia **and** evaluates the implementation of these reaction conditions against cost and safety
- 5 marks: explains, using Le Chatelier's principle, how the reaction conditions of temperature **and** pressure **and/or** concentration **and/or** catalyst can be used to increase the yield of ammonia **and** evaluates the implementation of these reaction conditions against cost and safety
- 4 marks: explains, using Le Chatelier's principle, how the reaction conditions of temperature **and** pressure **or** concentration **or** catalyst can be used to increase the yield of ammonia **and** attempts to evaluates the implementation of these reaction conditions against cost **and/or** safety
- 3 marks: explains how the reaction conditions of temperature **and** pressure **or** concentration **or** catalyst can be used to increase the yield of ammonia **and** attempts to evaluate the implementation of these reaction conditions against cost **and/or** safety
- 2 marks: explains how the reaction conditions of temperature **or** pressure **or** concentration **or** catalyst can be used to increase the yield of ammonia while addressing cost **or** safety
- 1 mark: provides some relevant information

Question 32

Because the white compound was insoluble, it cannot be a nitrate, sulfate or any group 17 compound (fluoride, chloride, bromide, iodide). It also cannot contain the cations of group 1 (Li^+, Na^+, K^+).

The white solid reacted with nitric acid; therefore, it must be a carbonate:

$$CO_3^{2-}(s) + 2H^+(aq) \rightarrow H_2O(l) + CO_2(g)$$

The resulting liquid would have to be a nitrate (from the nitric acid), so further testing would provide the identity of the cation.

As a precipitate formed with sodium sulfate solution and all sulfates are soluble except barium, lead and calcium, the anion must be one of these.

If it was lead, a precipitate would have formed when added to sodium chloride solution (which it did not).

If it were calcium, a precipitate would have formed when added to sodium hydroxide solution (which it did not).

Therefore, the insoluble white ionic compound was barium carbonate.

$$Ba^{2+}(aq) + SO_4^{2-}(aq) \rightarrow BaSO_4(s)$$

- 4 marks: identifies compound as barium carbonate with detailed reasoning **and** relevant ionic equations
- 3 marks: identifies compound as barium carbonate and calcium carbonate (cannot distinguish the two) **and** provides reasoning **and** relevant ionic equations
- 2 marks: makes steps towards correctly identifying compound **and** includes an ionic equation
- 1 mark: provides some relevant information

Question 33

a   Low temperature, yeast catalyst, anaerobic conditions

b   1. Use a mortar and pestle to crush 100 g of grapes.

    2. Add crushed grapes to a conical flask, along with 50 mL deionised water.

    3. Bloom 10 g of yeast with 20 mL of water in a 50 mL beaker.

    4. Add bloomed yeast to the conical flask and stir with a glass rod.

    5. Stopper the conical flask to ensure anaerobic conditions. Attach a plastic hose to allow carbon dioxide to be released.

    6. Place the conical flask into a water bath set at 37°C for 1 week.

    7. Filter the contents of the conical flask, collecting the filtrate into a second conical flask.

    8. Stopper the conical flask to ensure anaerobic conditions. Attach a plastic hose to allow carbon dioxide to escape.

- 3 marks: outlines a valid fermentation method for an alcohol that includes all conditions required, using school laboratory equipment
- 2 marks: outlines a fermentation method (sugar) that includes all conditions required, using school laboratory equipment
- 1 mark: outlines a fermentation method

c Iodine is a brown chemical. When mixed with ethanol in the presence of sodium hydroxide, the ethanol reacts in three steps to form a carboxylic acid and iodoform. The formation of the iodoform removes iodine from solution; therefore, in the presence of ethanol, the solution turns colourless. If methanol was the most prevalent alcohol, the solution would remain brown because halogenation would not occur.

- 3 marks: links the formation of iodoform with the removal of colour as an indicator of ethanol **and** links the lack of halogenation (colour change) with methanol
- 2 marks: links the formation of iodoform with the removal of colour as an indicator of ethanol **or** links the formation of iodoform with the presence of ethanol and the inability of methanol to produce iodoform
- 1 mark: provides some relevant information

Question 34

As the number of carbons increases, the boiling points of carboxylic acids, amines and alkanes increase. As carbon chain length increases, so does the strength of the collective intermolecular forces; therefore, increasing boiling point.

Carboxylic acids have higher boiling points than amines and alkanes. Alkanes are non-polar molecules and the only force of attraction between molecules is weak dispersion forces. Primary amines can form hydrogen bonds, which are strong intermolecular forces, between adjacent molecules because of the presence of the polar $-NH_2$ group. Carboxylic acids have the highest boiling points because the $C=O$ part of the functional group will form dipole–dipole forces due to its high polarity and the $-OH$ part of the functional group will form hydrogen bonds.

- 3 marks: links trend between carbon chain length to collective dispersion forces **and** links functional group trend to the different intermolecular forces involved with each functional group
- 2 marks: links trend between carbon chain length to collective dispersion forces **or** links functional group trend to the different intermolecular forces involved with each functional group
- 1 mark: identifies the trend between carbon chain length, as well as carboxylic acids > amines > alkanes

Question 35

$$n(C) = n(CO_2) = \frac{0.414\,22}{24.79} = 0.016\,709\ldots \text{ mol}$$

$$n(H) = 2 \times n(H_2O) = 2 \times (0.300\,616)/((2 \times 1.008) + (16.00)) = 0.033\,721\ldots \text{ mol}$$

$$m(O) = m(\text{compound}) - m(\text{carbon} + \text{hydrogen})$$

$$m(O) = 0.502 - ((0.016\,709\ldots \times 12.01) + (0.033\,721\ldots \times 1.008)) = 0.267\,334\ldots \text{ g}$$

$$n(O) = \frac{m(O)}{MM(O)} = \frac{0.267\,656}{16.00} = 0.016\,7083\ldots \text{ mol}$$

	C	H	O
n	0.016 709… mol	0.033 721… mol	0.0167 083 mol
Divide by the smallest	1	2	1

Empirical formula is $(CH_2O)_n$

Mass spectrum has parent ion peak at m/z 60

Therefore, $n \times (12.01 + (2 \times 1.008) + 16.00) = 60$

$$n \times 30 = 60$$

$$n = 2$$

Molecular formula is $C_2H_4O_2$

^{13}C NMR spectrum shows two peaks due to carbon environments.

^{1}H NMR spectrum shows two hydrogen environments and since there is no splitting pattern observed, there are no hydrogens on neighbouring carbons.

Infrared shows a very broad band at about 3000 cm^{-1} due to acid –O–H vibrations and a band at 1680–1750 cm^{-1} due to C=O stretching. There must be a carboxylic acid group present.

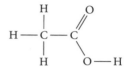

The ^{13}C peak at about 179 ppm confirms that the C is part of a carboxylic acid.

Therefore, the compound is ethanoic acid with the structural formula shown.

- 6 marks: calculates empirical formula **and** calculates molecular formula by referring to parent ion (*m/z* 60) in mass spectrum **and** draws correct structural formula with reference to ^{13}C and ^{1}H NMR and infrared spectra
- 5 marks: calculates empirical formula **and** calculates molecular formula by referring to parent ion (*m/z* 60) in mass spectrum **and** draws correct structural formula with reference to at least two other spectra
- 3–4 marks: calculates empirical formula **and** calculates molecular formula by referring to parent ion (*m/z* 60) in mass spectrum **and** draws correct structural formula with reference to one other spectrum
- 1–2 marks: calculates empirical formula **and** calculates molecular formula by referring to parent ion in mass spectrum **and** draws correct structural formula

Practice HSC exam 2

Multiple-choice solutions

Question 1

A Addition

Chlorine is a halogen and undergoes the same addition reaction at a double bond as bromine.

Question 2

D I and III only

Arrhenius acids are hydrogen-containing compounds that produce H^{+} ions when dissociated in water. Arrhenius bases are hydroxide compounds that produce OH^{-} ions on dissociation in water. Reactions II and IV are modelling Brønsted–Lowry proton donation.

Question 3

B

A volumetric pipette is used to deliver an aliquot in a titration.

Question 4

B Ethanal

Ethanal is produced when ethanol reacts with acidified potassium dichromate and it has a lower boiling point than water. Ethanoic acid is produced if the mixture is refluxed.

Question 5

C The mixture becomes a lighter colour when temperature is decreased.

The reaction is endothermic, so a decrease in temperature will cause the equilibrium to shift to the left and result in a lighter colour.

Question 6

B   Methanol and butanoic acid

It is by looking at the structure of the ester that the reactants can be identified. The carboxylic acid is drawn first with the alcohol replacing the hydrogen atom on the acid. In this case, we have butanoic acid (four carbons before the –O– with a methyl group (methanol) replacing the hydrogen atom on the acid.

When naming an ester, put the alcohol first (methyl) and then the carboxylic acid, changing the suffix to '-oate' (butanoate).

Question 7

D

The hydrophobic tail positions itself in the oil, as far from the water as possible. The hydrophilic head remains on the outside of the oil, forming a micelle.

Question 8

B $C_6H_5NH_2(l)$ and $C_6H_5NH^-(aq)$

Acids donate protons, and bases accept protons. When $C_6H_5NH_2$ donates a proton to water, it becomes $C_6H_5NH^-$. This conjugate can accept the proton back from water, making it a conjugate base. H_3O^+ is the conjugate acid because H_2O is acting as a base.

Question 9

B $H_2PO_4^-$

The conjugate acid has one more proton.

Question 10

C 6.41×10^{-6}

$K_{sp} = 0.0117 \times (2 \times 0.117)^2 = 6.41 \times 10^{-6}$

Question 11

C   Aqueous sodium hydroxide was added to the system, which became a lighter colour.

Addition of NaOH(aq) precipitates Fe^{3+} ions, causing an immediate decrease in the concentration of Fe^{3+} ions, and the equilibrium shifts to the left, resulting in a lighter colour.

Question 12

C

Condensation polymers are formed when two monomers bond, producing a small molecule such as H_2O as a by-product. If A represents H, and B represents OH, these letters would no longer be present as H_2O is formed, leaving the shapes to connect and form the polymer. When repeated, this would remove all the letters, leaving just the shapes bonded together.

Question 13

C   Cresolphthalein (pH range 8.2–9.8)

The end point of a titration (when the indicator changes colour) should be as close as possible to the equivalence point. Based on the graph, this occurs at about pH 9.

Question 14

D No. $Q < K_{sp}$

$n(Ba^{2+}) = cV = 0.0500 \times 0.100 = 0.005 \, mol$

$c(Ba^{2+}) = \dfrac{0.005}{0.200} = 0.025 \, mol \, L^{-1}$

$n(OH^-) = cV = 0.100 \times 0.100 = 0.01 \, mol$

$c(OH^-) = \dfrac{0.01}{0.200} = 0.05 \, mol \, L^{-1}$

$Q_{sp} = 0.025 \times 0.05^2 = 6.25 \times 10^{-5}$

$K_{sp} = 2.55 \times 10^{-4}$

Therefore $Q_{sp} < K_{sp}$

Question 15

B Calcium hydroxide and copper(II) sulfate

Two precipitates would form: $Cu(OH)_2$ and $CaSO_4$.

Question 16

A 2

There are two hydrogen environments, as shown in the diagram.

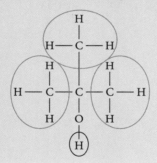

Question 17

D Propan-2-ol

The peaks are labelled as shown.

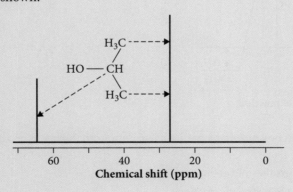

The chemical shift below 60 is attributed to a carbon attached to a hydroxyl group.

Question 18

B Nitric acid

Nitric acid will produce a colourless, odourless gas when added to carbonate ions but no visible reaction with Cl^-. Lead(II) nitrate (**A**) and silver nitrate (**C**) will form precipitates with both ions and there is no visible reaction with sodium hydroxide (**D**).

Question 19

C Copper

The solution is likely to contain copper ions, which are blue in colour. The spectrum shows the complementary yellow light being absorbed and no absorption in the blue area.

Question 20

D 93.1 mg

$c(MnO_4^-) = 0.20\,mmol\,L^{-1}$ (from graph)

In 250 mL, $0.250 \times \dfrac{0.20}{1000} = 0.000\,05\,mol$

In 1.00 L, $40 \times 0.000\,05 = 0.002\,mol$

From the equation $n(Fe^{2+}) = 5 \times n(MnO_4^-) = 5 \times 0.002 = 0.01\,mol$

$m(Fe^{2+}) = 0.01 \times 55.85 = 0.5585$ g

Since there were 6 tablets, mass per tablet $= \dfrac{0.5585}{6} = 0.0931\,g$

$0.0931 \times 1000 = 93.1\,mg$

Short-answer solutions

Question 21

a $K = \dfrac{[NH_3]^2}{[N_2][H_2]^3}$

b

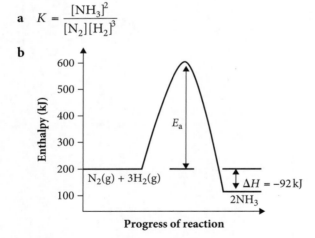

- 3 marks: correctly sketches an energy profile using data from the question with E_a, ΔH and products labelled
- 2 marks: sketches an exothermic energy profile with labels
- 1 mark: provides some relevant information

c   At time A, ammonia was removed from the system as seen by the sharp decline in concentration. A reduction in the concentration of ammonia would cause the rate of the reverse reaction to decrease, while the rate of the forward reaction would increase. This is seen by the decline in concentration of both nitrogen and hydrogen, while the concentration of ammonia increased until equilibrium was restored. At time B, equilibrium was restored and the rates of the forward and reverse reactions were again equal.

- **3 marks**: identifies that ammonia was removed from the system **and** uses collision theory to predict the rate of reaction **and** makes direct reference to the shape of the curves
- **2 marks**: identifies that ammonia was removed from the system **and** attempts to use collision theory to explain an increase in the rate of the forward reaction
- **1 mark**: identifies that ammonia was removed from the system

d

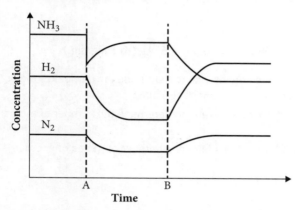

- **2 marks**: correctly draws the increase and decrease in concentrations, using stoichiometric ratio, until equilibrium is restored at the same point
- **1 mark**: correctly draws the increase and decrease in concentrations until equilibrium is restored

Question 22

a Two of: available in very high purity, very low reactivity with CO_2 and/or O_2, very low reactivity with moisture in the air, predictable reactivity

b Measuring cylinder, pipette, burette

c n(perchloric/ethanoic acid solution) in 15.3 mL = $c \times V$ = 0.0583 × 0.0153 = 8.9199 × 10^{-4} mol

Molar ratio of acid : nicotine = 2 : 1, therefore

n(nicotine) in 20 mL = $\dfrac{2}{8.9199 \times 10^{-4}}$ = 4.46 × 10^{-4} mol

n(nicotine) in 100 mL = 4.46 × 10^{-4} × 5 = 2.23 × 10^{-3} mol

MM(nicotine) = (10 × 12.01) + (14 × 1.008) + (2 × 14.01) = 162.232

m(nicotine) = $MM \times n$ = 162.232 × 2.23 × 10^{-3} = 0.3618 g × 1000 = 361.8 mg

m(nicotine) per capsule = $\dfrac{361.8}{25}$ = 14.5 mg

- **5 marks**: calculates the number of moles of perchloric/ethanoic acid solution used **and** uses molar ratio to determine n(nicotine) in 100 mL **and** calculates MM of nicotine **and** determines the mass of nicotine in each capsule in mg
- **4 marks**: calculates the number of moles of perchloric/ethanoic acid solution used **and** uses molar ratio to determine n(nicotine) in 100 mL **and** calculates MM of nicotine **and** determines the mass of nicotine in solution
- **3 marks**: calculates the number of moles of perchloric/ethanoic acid solution used **and** determines n(nicotine) in 20 mL **and** calculates MM of nicotine **and/or** determines the mass of nicotine in solution
- **2 marks**: calculates the number of moles of perchloric/ethanoic acid solution used **and** determines n(nicotine) in 20 mL
- **1 mark**: calculates the number of moles of perchloric/ethanoic acid solution used

Question 23

a $C_{12}H_{22}O_{11}(aq) + H_2O(l) \rightarrow 4C_2H_5OH(aq) + 4CO_2(g)$

$$n(\text{sucrose}) = \frac{m}{MM} = \frac{1250}{342} = 3.655 \, \text{mol}$$

Molar ratio sucrose : ethanol = 1 : 4, therefore

$n(\text{ethanol}) = 4 \times 3.655 = 14.6199 \, \text{mol}$

$m(\text{ethanol}) = n \times MM = 14.6199 \times (2 \times 12.01 + 6 \times 1.008 + 1 \times 16.00) = 674 \, \text{g (to 3 sig. fig.)}$

- 3 marks: provides balanced chemical equation **and** uses molar ratio to determine the number of moles of ethanol **and** correctly calculates the mass of ethanol to 3 significant figures
- 2 marks: provides balanced chemical equation **and** uses molar ratio to determine the number of moles of ethanol
- 1 mark: provides balanced chemical equation **or** correctly calculates the number of moles of sucrose

b $C_2H_4(g) + H_2O(l) \xrightarrow{\text{catalyst}} C_2H_5OH(g)$

c

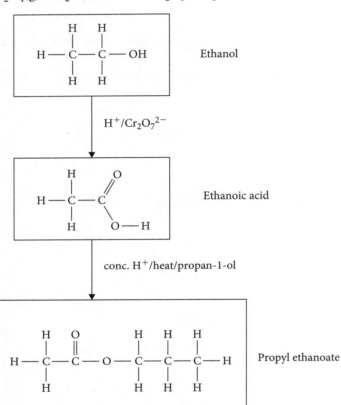

- 4 marks: provides correct structural formulae of ethanoic acid and propyl ethanoate **and** correctly names ethanoic acid **and** correctly names all reactants and conditions
- 3 marks: provides correct structural formulae of ethanoic acid and propyl ethanoate **and** correctly names ethanoic acid **and** correctly names one reactant/conditions required
- 2 marks: provides correct structural formulae of ethanoic acid and propyl ethanoate **and** correctly names ethanoic acid **or** correctly names one reactant/conditions required
- 1 mark: provides correct structural formula of ethanoic acid **or** propyl ethanoate

Question 24

a

Propanal   Propanone

- 2 marks: correctly draws propanal **and** propanone
- 1 mark: correctly draws propanal **or** propanone

b **i** CHO^+ or $CH_3CH_2^+$

ii The $CH_3CH_2^+/CHO^+$ fragment at m/z 29 indicates that propanal produced this spectrum. Neither prop-2-en-1-ol nor propanone would produce this fragment as their most intense fragment.

- 3 marks: correctly identifies propanal **and** justifies selection by fragmentation at m/z 29 **and** states other isomers would not produce this fragment
- 2 marks: correctly identifies propanal **and** justifies selection by fragmentation at m/z 29
- 1 mark: correctly identifies propanal

c Propanone is symmetrical. Both carbon atoms at the ends of the molecule are the same; therefore, there are two environments that correspond to the two peaks in the ^{13}C NMR spectrum. The ^1H NMR spectrum shows one peak, meaning all hydrogens are equivalent. A peak at 2.2 ppm also corresponds to a ketone.

Prop-2-en-1-ol and propanal both have three carbon environments and multiple hydrogen environments.

- 3 marks: correctly identifies propanone **and** justifies selection by referring to the peaks and therefore environments on the ^{13}C NMR spectrum **and** justifies selection by referring to the peak and therefore environment on the ^1H NMR spectrum
- 2 marks: correctly identifies propanone **and** justifies selection by referring to the peaks and therefore environments on the ^{13}C NMR spectrum **or** justifies selection by referring to the peak and therefore environment on the ^1H NMR spectrum
- 1 mark: correctly identifies propanone

Question 25

a

Label	Liquid
W	Ethanoic acid
X	Ethanol
Y	2-Methylpropan-2-ol
Z	Cyclohexene

- 2 marks: identifies all four liquids
- 1 mark: identifies two or three liquids

b **i** When the colourless liquid was added to the purple potassium permanganate, the colour faded.

- 2 marks: states a complete observation
- 1 mark: states a colour

ii

Ethanoic acid   Cyclohexane-1,2-diol

- 2 marks: gives IUPAC name **and** correctly draws the structural formula of the product formed
- 1 mark: gives IUPAC name **or** correctly draws the structural formula of the product formed

Question 26

Initially, the volume of the chamber increases, providing more space for the atoms to spread out. This is seen by the mixture first turning a light brown colour.

An increase in volume (decrease in pressure), at constant temperature, will shift the equilibrium to the side where there are more moles of gas. In this case, the equilibrium will shift to the left, forming more NO_2. Hence the mixture becomes darker brown.

- 3 marks: justifies the two colour changes
- 2 marks: justifies one colour change
- 1 mark: provides some relevant information

Question 27

a It is a control and is used to observe whether there is any decrease in mass without the yeast. Since there is minimal loss in mass it must be due to evaporation of water.

b The water bath is used to maintain a constant temperature of 37°C.

- 2 marks: states the need to maintain a constant temperature **and** identifies the temperature
- 1 mark: states the need to maintain a constant temperature **or** identifies the temperature

c The flask with the limewater is used so the carbon dioxide gas can be bubbled directly into it without releasing it to the atmosphere because of the need to maintain anaerobic conditions. The limewater turns milky then colourless again when carbon dioxide is bubbled through it.

- 2 marks: explains the purpose of the limewater flask
- 1 mark: states the purpose of the limewater flask

d Equation: $C_6H_{12}O_6(aq) \rightarrow 2CH_3CH_2OH(aq) + 2CO_2(g)$

Mass decrease = mass of CO_2 = 235.16 − 230.18 = 4.98 g

$$n(CO_2) = \frac{m}{MM} = \frac{4.98}{12.01 + (2 \times 16.00)} = 0.1131\ldots \text{mol}$$

$n(\text{ethanol}) = n(CO_2) = 0.1131\ldots \text{mol}$

$m(\text{ethanol}) = n \times MM = 0.1131\ldots \times 46.068 = 5.2129\ldots \text{g}$

$n(\text{glucose}) = \frac{1}{2}n(CO_2) = \frac{1}{2} \times 0.1131\ldots = 0.05657\ldots \text{mol}$

$m(\text{glucose}) = n \times MM = 0.05657\ldots \times ((6 \times 12.01) + (12 \times 1.008) + (6 \times 16.00)) = 10.19 \text{g}$

Theoretical yield:

$$n(\text{glucose}) = \frac{67.93}{MM} = \frac{67.93}{180.156} = 0.377\,06\ldots \text{mol}$$

$$n(\text{ethanol}) = 2 \times n(\text{glucose}) = 2 \times 0.377\,06\ldots \text{mol}$$

$$m(\text{ethanol}) = n \times MM = 2 \times 0.377\,06\ldots \times 46.068 = 34.7409\ldots\text{g}$$

$$\% \text{ yield of ethanol} = \frac{5.2129\ldots}{34.7409\ldots} \times 100 = 15.00\% \text{ (to 4 sig. fig.)}$$

- **5 marks:** writes correct balanced equation with states **and** for experimental results **and** calculates number of moles of CO_2 **and** states number of moles of ethanol **and** calculates mass of ethanol **and** calculates number of moles of glucose and mass of glucose **and** for theoretical results. Calculates number of moles of glucose **and** calculates number of moles and mass of ethanol **and** calculates % yield of ethanol to correct significant figures

- **3–4 marks:** writes correct balanced equation with states **and** for experimental results **and** calculates number of moles of CO_2 **and** states number of moles of ethanol **and** calculates mass of ethanol **and** calculates number of moles of glucose and mass of glucose **and** for theoretical results. Calculates number of moles of glucose **and** calculates number of moles and mass of ethanol **and** calculates % yield of ethanol (incorrect significant figures) **or** writes correct balanced equation with states **and** does two correct steps for experimental results **and** does two correct steps for theoretical results

- **1–2 marks:** writes correct balanced equation with states **and/or** calculates number of moles of CO_2 **and/or** states number of moles of ethanol **and/or** calculates mass of ethanol **and/or** calculates number of moles of glucose and mass of glucose **and** for theoretical results **and** calculates number of moles of glucose **and/or** calculates number of moles and mass of ethanol **and/or** calculates % yield of ethanol (incorrect significant figures)

Question 28

From graph:
For $A = 0.525$
$$y = 0.0659x$$

Species	Fe^{3+}	SCN^-	$Fe(SCN)^{2+}$
I	$0.1 \times 0.02 = 2.0 \times 10^{-3}\,\text{mol}$	$0.1 \times 0.02 = 2.0 \times 10^{-3}\,\text{mol}$	$0\,\text{mol}$
C	$-4.779\,96 \times 10^{-4}\,\text{mol}$	$-4.779\,96 \times 10^{-4}\,\text{mol}$	$+4.779\,96 \times 10^{-4}\,\text{mol}$
E	$2.0 \times 10^{-3} - 4.779\,96 \times 10^{-4}$ $= 1.522 \times 10^{-3}\,\text{mol}$	$2.0 \times 10^{-3} - 4.779\,96 \times 10^{-4}$ $= 1.522 \times 10^{-3}\,\text{mol}$	$n = cV = 7.9666\ldots \times 10^{-3} \times 0.06$ $= 4.779\,96 \times 10^{-4}\,\text{mol}$
E concentration	$c = \dfrac{n}{V}$ $= 2.5366\ldots \times 10^{-2}\,\text{mol L}^{-1}$	$c = \dfrac{n}{V}$ $= 2.5366\ldots \times 10^{-2}\,\text{mol L}^{-1}$	From graph: For $A = 0.525$ $y = 0.0659x$ $= \dfrac{0.525}{0.0659}$ $= 7.9666\ldots \text{mmol mL}^{-1}$ $= \dfrac{7.996}{10^6}$ $= 7.9666\ldots \times 10^{-6} \times 1000\,\text{mol L}^{-1}$ $= 7.9666\ldots \times 10^{-3}\,\text{mol L}^{-1}$

$$K_{eq} = \frac{7.9666 \times 10^{-3}}{(2.5367 \times 10^{-2})(2.5367 \times 10^{-2})} = 12.38 = 12.4 \text{ (to 3 sig. fig.)}$$

- **5 marks:** provides correct K_{eq} value and to 3 significant figures
- **3–4 marks:** provides correct K_{eq} value **or** calculates concentration of $Fe(SCN)^{2+}$ from gradient/graph **and** calculates number of moles of $Fe(SCN)^{2+}$ at equilibrium **and/or** calculates the change in number of moles of Fe^{3+} **and/or** calculates the change in number of moles of SCN^- **and/or** calculates the concentration of Fe^{3+} **and/or** calculates the concentration of SCN^-
- **1–2 marks:** calculates concentration of $Fe(SCN)^{2+}$ from gradient/graph

Question 29

a $n(\text{propanol}) = \dfrac{m}{MM} = \dfrac{0.632}{60.094} = 0.010\,517\,\text{mol}$

$q = mC\Delta T = 105 \times 4.18 \times (65 - 16.8) = \dfrac{21\,154.98}{1000} = 21.15\,\text{kJ}$

$\Delta H = \dfrac{q}{n} = \dfrac{21.15}{0.010\,52} = -2011.5308 = -2.01 \times 10^3\,\text{kJ mol}^{-1}$ (to 3 sig. fig.)

- 4 marks: calculates moles of propanol **and** calculates q **and** calculates ΔH to 3 significant figures
- 3 marks: calculates moles of propanol **and** calculates q **and** calculates ΔH
- 2 marks: calculates moles of propanol **and** calculates q
- 1 mark: provides some relevant information

b The experimental value for the combustion of propanol is less than the theoretical value primarily because of heat loss to the environment. Using an open, uninsulated aluminium beaker allows heat to escape. The spirit burner is also a considerable distance from the aluminium beaker so heat is lost before it can enter the water.

- 2 marks: identifies that experimental value is less than theoretical value **and** provides reasoning specific to the experimental procedure
- 1 mark: provides some relevant information

Question 30

a The curve steadily falls as OH^- ions are added because they react and remove the highly conductive H^+ ions from the solution. At the equivalence point, conductivity is lowest because the solution contains only NH_4^+ and NO_3^- ions, which are much less mobile than the H^+ or OH^- ions. The curve then plateaus because ammonium hydroxide is a weak base and its dissociation rate is slow, providing few OH^- ions.

- 3 marks: explains the shape of the curve in terms of ions present as it falls and as it plateaus
- 2 marks: describes the trend in the curve and provides an explanation
- 1 mark: provides some relevant information

b $HNO_3(aq) + NH_4OH(aq) \rightarrow NH_4NO_3(aq) + H_2O(l)$

$n(HNO_3) = cV = 1.03 \times 10^{-4} \times 0.2 = 0.000\,0206\,\text{mol}$

Mole ratio $HNO_3 : NH_4OH = 1 : 1$; therefore, $n(NH_4OH) = 0.000\,0206\,\text{mol}$

$c(NH_4OH) = \dfrac{n}{V} = \dfrac{0.000\,0206}{0.017\,15} = 1.20 \times 10^{-3}\,\text{mol L}^{-1}$ (to 3 sig. fig.)

- 4 marks: provides correct balanced chemical equation with states **and** calculates the number of moles of HNO_3 and NH_4OH **and** calculates concentration of NH_4OH to 3 significant figures
- 3 marks: provides correct balanced chemical equation **and** calculates the number of moles of HNO_3 and NH_4OH **and** calculates concentration of NH_4OH
- 2 marks: provides balanced chemical equation **and** calculates the number of moles of HNO_3 and NH_4OH
- 1 mark: provides some relevant information

9780170465274

Question 31

$$2Al(OH)_3(s) + 3H_2SO_4(aq) \rightarrow Al_2(SO_4)_3(aq) + 6H_2O(l)$$

$$n(Al(OH)_3) = \frac{m}{MM} = \frac{1.17}{78.004} = 0.0150 \, mol$$

$$n(H_2SO_4) = cV = 0.125 \times 0.4 = 0.05 \, mol$$

$$n(H_2SO_4) \text{ reacting with } Al(OH)_3 = \frac{3}{2} \times 0.0150 = 0.0225 \, mol$$

$$n(H_2SO_4 \text{ in excess}) = 0.05 - 0.0225 = 0.0275 \, mol$$

$$c(H_2SO_4) = \frac{n}{V} = \frac{0.0275}{0.4} = 0.068\,75 \, mol \, L^{-1}$$

$$pH = -\log[0.068\,75] = 1.16$$

- **4 marks:** provides correctly balanced chemical equation **and** determines number of moles of H_2SO_4 **and** determines number of moles of H_2SO_4 in excess **and** calculates concentration of H_2SO_4 **and** correctly calculates pH of resultant solution
- **3 marks:** provides correctly balanced chemical equation **and** determines number of moles of $Al(OH)_3$ **and** determines number of moles of H_2SO_4 **and** makes steps to determine number of moles of H_2SO_4 in excess
- **2 marks:** provides correctly balanced chemical equation **and** determines number of moles of $Al(OH)_3$ **or** determines number of moles of H_2SO_4
- **1 mark:** provides some relevant information

Question 32 ©NESA 2020 MARKING GUIDELINES SII Q23

Sample answer:

The use of a catalyst in reactions 1 and 2 allows a faster rate at a lower temperature. This both increases the efficiency of the process and reduces energy consumption that makes the process more economical and ultimately less polluting.

Unused reactant gases are recycled after being separated from the reaction mixture in separator 1. This means that resources are not wasted – making the process more economical as less reactant needs to be purchased from suppliers.

Markets have been accessed for the major product (ethane-1,2-diol) – without a market the industrial process is not economically viable – the location of these markets has been determined and suitable transport arranged – the industrial plant has probably been located near a major port, rail or road network to facilitate economical and rapid transport to markets.

Answers could include:

- By-products are also produced rather than wastes to be disposed of. These by-products are sent to markets, which indicates that all the products of the reaction have a purpose. This makes the entire process more economical and less wasteful (potentially 100% atom economy).

- Access to reactants – the plant would be located so that ethylene and oxygen gases would be easily accessible – perhaps near a petrochemical plant or road/rail network so that these resources could be transported easily and cheaply.

  - **4 marks:** explains three relevant factors and makes specific reference to the flow chart
  - **3 marks:** explains two relevant factors with some reference to the flow chart **or** explains one relevant factor and outlines two other relevant factors with some reference to the flow chart **or** explains three relevant factors without specific reference to the flow chart
  - **2 marks:** explains one relevant factor **or** outlines two relevant factors
  - **1 mark:** provides some relevant information

HIGHER SCHOOL CERTIFICATE EXAMINATION

Formulae sheet

$$n = \frac{m}{MM}$$

$$c = \frac{n}{V}$$

$$PV = nRT$$

$$q = mc\Delta T$$

$$\Delta G° = \Delta H° - T\Delta S°$$

$$pH = -\log_{10}[H^+]$$

$$pK_a = -\log_{10}[K_a^+]$$

$$A = \varepsilon lc = \log_{10}\frac{I_o}{I}$$

Avogadro constant, N_A	$6.022 \times 10^{23}\,\text{mol}^{-1}$
Volume of 1 mole ideal gas: at 100 kPa and	
at 0°C (273.15 K)	22.71 L
at 25°C (298.15 K)	24.79 L
Gas constant	$8.314\,\text{J}\,\text{mol}^{-1}\,\text{K}^{-1}$
Ionisation constant for water at 25°C (298.15 K), K_w	1.0×10^{-14}
Specific heat capacity of water	$4.18 \times 10^3\,\text{J}\,\text{kg}^{-1}\,\text{K}^{-1}$

Data sheet

Solubility constants at 25°C

Compound	K_{sp}	Compound	K_{sp}
Barium carbonate	2.58×10^{-9}	Lead(II) bromide	6.60×10^{-6}
Barium hydroxide	2.55×10^{-4}	Lead(II) chloride	1.70×10^{-5}
Barium phosphate	1.3×10^{-29}	Lead(II) iodide	9.8×10^{-9}
Barium sulfate	1.08×10^{-10}	Lead(II) carbonate	7.40×10^{-14}
Calcium carbonate	3.36×10^{-9}	Lead(II) hydroxide	1.43×10^{-15}
Calcium hydroxide	5.02×10^{-6}	Lead(II) phosphate	8.0×10^{-43}
Calcium phosphate	2.07×10^{-29}	Lead(II) sulfate	2.53×10^{-8}
Calcium sulfate	4.93×10^{-5}	Magnesium carbonate	6.82×10^{-6}
Copper(II) carbonate	1.4×10^{-10}	Magnesium hydroxide	5.61×10^{-12}
Copper(II) hydroxide	2.2×10^{-20}	Magnesium phosphate	1.04×10^{-24}
Copper(II) phosphate	1.40×10^{-37}	Silver bromide	5.35×10^{-13}
Iron(II) carbonate	3.13×10^{-11}	Silver chloride	1.77×10^{-10}
Iron(II) hydroxide	4.87×10^{-17}	Silver carbonate	8.46×10^{-12}
Iron(III) hydroxide	2.79×10^{-39}	Silver hydroxide	2.0×10^{-8}
Iron(III) phosphate	9.91×10^{-16}	Silver iodide	8.52×10^{-17}
		Silver phosphate	8.89×10^{-17}
		Silver sulfate	1.20×10^{-5}

Infrared absorption data

Bond	Wavenumber/cm^{-1}
N—H (amines)	3300–3500
O—H (alcohols)	3230–3550 (broad)
C—H	2850–3300
O—H (acids)	2500–3000 (very broad)
C≡N	2220–2260
C=O	1680–1750
C=C	1620–1680
C—O	1000–1300
C—C	750–1100

^{13}C NMR chemical shift data

Type of carbon	δ/ppm
—C—C—	5–40
R—C—Cl or Br	10–70
R—C—C— (C=O)	20–50
R—C—N	25–60
—C—O— alcohols, ethers or esters	50–90
C=C	90–150
R—C≡N	110–125
benzene ring	110–160
R—C=O esters or acids	160–185
R—C=O aldehydes or ketones	190–220

UV absorption

(This is not a definitive list and is approximate.)

Chromophore	λ_{max} (nm)
C—H	122
C—C	135
C=C	162

Chromophore	λ_{max} (nm)
C≡C	173 178 196 222
C—Cl	173
C—Br	208

Some standard potentials

$K^+ + e^-$	$\rightleftharpoons$	$K(s)$	$-2.94\,V$
$Ba^{2+} + 2e^-$	$\rightleftharpoons$	$Ba(s)$	$-2.91\,V$
$Ca^{2+} + 2e^-$	$\rightleftharpoons$	$Ca(s)$	$-2.87\,V$
$Na^+ + e^-$	$\rightleftharpoons$	$Na(s)$	$-2.71\,V$
$Mg^{2+} + 2e^-$	$\rightleftharpoons$	$Mg(s)$	$-2.36\,V$
$Al^{3+} + 3e^-$	$\rightleftharpoons$	$Al(s)$	$-1.68\,V$
$Mn^{2+} + 2e^-$	$\rightleftharpoons$	$Mn(s)$	$-1.18\,V$
$H_2O + e^-$	$\rightleftharpoons$	$\frac{1}{2}H_2(g) + OH^-$	$-0.83\,V$
$Zn^{2+} + 2e^-$	$\rightleftharpoons$	$Zn(s)$	$-0.76\,V$
$Fe^{2+} + 2e^-$	$\rightleftharpoons$	$Fe(s)$	$-0.44\,V$
$Ni^{2+} + 2e^-$	$\rightleftharpoons$	$Ni(s)$	$-0.24\,V$
$Sn^{2+} + 2e^-$	$\rightleftharpoons$	$Sn(s)$	$-0.14\,V$
$Pb^{2+} + 2e^-$	$\rightleftharpoons$	$Pb(s)$	$-0.13\,V$
$H^+ + e^-$	$\rightleftharpoons$	$\frac{1}{2}H_2(g)$	$0.00\,V$
$SO_4^{2-} + 4H^+ + 2e^-$	$\rightleftharpoons$	$SO_2(aq) + 2H_2O$	$0.16\,V$
$Cu^{2+} + 2e^-$	$\rightleftharpoons$	$Cu(s)$	$0.34\,V$
$\frac{1}{2}O_2(g) + H_2O + 2e^-$	$\rightleftharpoons$	$2OH^-$	$0.40\,V$
$Cu^+ + e^-$	$\rightleftharpoons$	$Cu(s)$	$0.52\,V$
$\frac{1}{2}I_2(s) + e^-$	$\rightleftharpoons$	I^-	$0.54\,V$
$\frac{1}{2}I_2(aq) + e^-$	$\rightleftharpoons$	I^-	$0.62\,V$
$Fe^{3+} + e^-$	$\rightleftharpoons$	Fe^{2+}	$0.77\,V$
$Ag^+ + e^-$	$\rightleftharpoons$	$Ag(s)$	$0.80\,V$
$\frac{1}{2}Br_2(l) + e^-$	$\rightleftharpoons$	Br^-	$1.08\,V$
$\frac{1}{2}Br_2(aq) + e^-$	$\rightleftharpoons$	Br^-	$1.10\,V$
$\frac{1}{2}O_2(g) + 2H^+ + 2e^-$	$\rightleftharpoons$	H_2O	$1.23\,V$
$\frac{1}{2}Cl_2(g) + e^-$	$\rightleftharpoons$	Cl^-	$1.36\,V$
$\frac{1}{2}Cr_2O_7^{2-} + 7H^+ + 3e^-$	$\rightleftharpoons$	$Cr^{3+} + \frac{7}{2}H_2O$	$1.36\,V$
$\frac{1}{2}Cl_2(aq) + e^-$	$\rightleftharpoons$	Cl^-	$1.40\,V$
$MnO_4^- + 8H^+ + 5e^-$	$\rightleftharpoons$	$Mn^{2+} + 4H_2O$	$1.51\,V$
$\frac{1}{2}F_2(g) + e^-$	$\rightleftharpoons$	F^-	$2.89\,V$

Aylward and Findlay, *SI Chemical Data* (5th Edition) is the principal source of data for the standard potentials. Some data may have been modified for examination purposes.

Periodic table of the elements

KEY

79
Au
197.0
Gold

Atomic Number
Symbol
Standard Atomic Weight
Name

1 H 1.008 Hydrogen																	2 He 4.003 Helium
3 Li 6.941 Lithium	4 Be 9.012 Beryllium											5 B 10.81 Boron	6 C 12.01 Carbon	7 N 14.01 Nitrogen	8 O 16.00 Oxygen	9 F 19.00 Fluorine	10 Ne 20.18 Neon
11 Na 22.99 Sodium	12 Mg 24.31 Magnesium											13 Al 26.98 Aluminium	14 Si 28.09 Silicon	15 P 30.97 Phosphorus	16 S 32.07 Sulfur	17 Cl 35.45 Chlorine	18 Ar 39.95 Argon
19 K 39.10 Potassium	20 Ca 40.08 Calcium	21 Sc 44.96 Scandium	22 Ti 47.87 Titanium	23 V 50.94 Vanadium	24 Cr 52.00 Chromium	25 Mn 54.94 Manganese	26 Fe 55.85 Iron	27 Co 58.93 Cobalt	28 Ni 58.69 Nickel	29 Cu 63.55 Copper	30 Zn 65.38 Zinc	31 Ga 69.72 Gallium	32 Ge 72.64 Germanium	33 As 74.92 Arsenic	34 Se 78.96 Selenium	35 Br 79.90 Bromine	36 Kr 83.80 Krypton
37 Rb 85.47 Rubidium	38 Sr 87.61 Strontium	39 Y 88.91 Yttrium	40 Zr 91.22 Zirconium	41 Nb 92.91 Niobium	42 Mo 95.96 Molybdenum	43 Tc Technetium	44 Ru 101.1 Ruthenium	45 Rh 102.9 Rhodium	46 Pd 106.4 Palladium	47 Ag 107.9 Silver	48 Cd 112.4 Cadmium	49 In 114.8 Indium	50 Sn 118.7 Tin	51 Sb 121.8 Antimony	52 Te 127.6 Tellurium	53 I 126.9 Iodine	54 Xe 131.3 Xenon
55 Cs 132.9 Caesium	56 Ba 137.3 Barium	57–71 Lanthanoids	72 Hf 178.5 Hafnium	73 Ta 180.9 Tantalum	74 W 183.9 Tungsten	75 Re 186.2 Rhenium	76 Os 190.2 Osmium	77 Ir 192.2 Iridium	78 Pt 195.1 Platinum	79 Au 197.0 Gold	80 Hg 200.6 Mercury	81 Tl 204.4 Thallium	82 Pb 207.2 Lead	83 Bi 209.0 Bismuth	84 Po Polonium	85 At Astatine	86 Rn Radon
87 Fr Francium	88 Ra Radium	89–103 Actinoids	104 Rf Rutherfordium	105 Db Dubnium	106 Sg Seaborgium	107 Bh Bohrium	108 Hs Hassium	109 Mt Meitnerium	110 Ds Darmstadtium	111 Rg Roentgenium	112 Cn Copernicium	113 Nh Nihonium	114 Fl Flerovium	115 Mc Moscovium	116 Lv Livermorium	117 Ts Tennessine	118 Og Oganesson

Lanthanoids

57 La 138.9 Lanthanum	58 Ce 140.1 Cerium	59 Pr 140.9 Praseodymium	60 Nd 144.2 Neodymium	61 Pm Promethium	62 Sm 150.4 Samarium	63 Eu 152.0 Europium	64 Gd 157.3 Gadolinium	65 Tb 158.9 Terbium	66 Dy 162.5 Dysprosium	67 Ho 164.9 Holmium	68 Er 167.3 Erbium	69 Tm 168.9 Thulium	70 Yb 173.1 Ytterbium	71 Lu 175.0 Lutetium

Actinoids

89 Ac Actinium	90 Th 232.0 Thorium	91 Pa 231.0 Protactinium	92 U 238.0 Uranium	93 Np Neptunium	94 Pu Plutonium	95 Am Americium	96 Cm Curium	97 Bk Berkelium	98 Cf Californium	99 Es Einsteinium	100 Fm Fermium	101 Md Mendelevium	102 No Nobelium	103 Lr Lawrencium

Standard atomic weights are abridged to four significant figures.

Elements with no reported values in the tables have no stable nuclides.

Information on elements with atomic numbers 113 and above is sourced from the International Union of Pure and Applied Chemistry Periodic Table of Elements (November 2016 version).

The International Union of Pure and Applied Chemistry Periodic Table of the Elements (February 2010 version) is the principal source of all other data. Some data may have been modified.

2021 Higher School Certificate Examination © copyright 2021, NSW Education Standards Authority